JEPPESEN

Sanderson Train

A&P TECHNICIAN

POWERPLANT

WORKBOOK

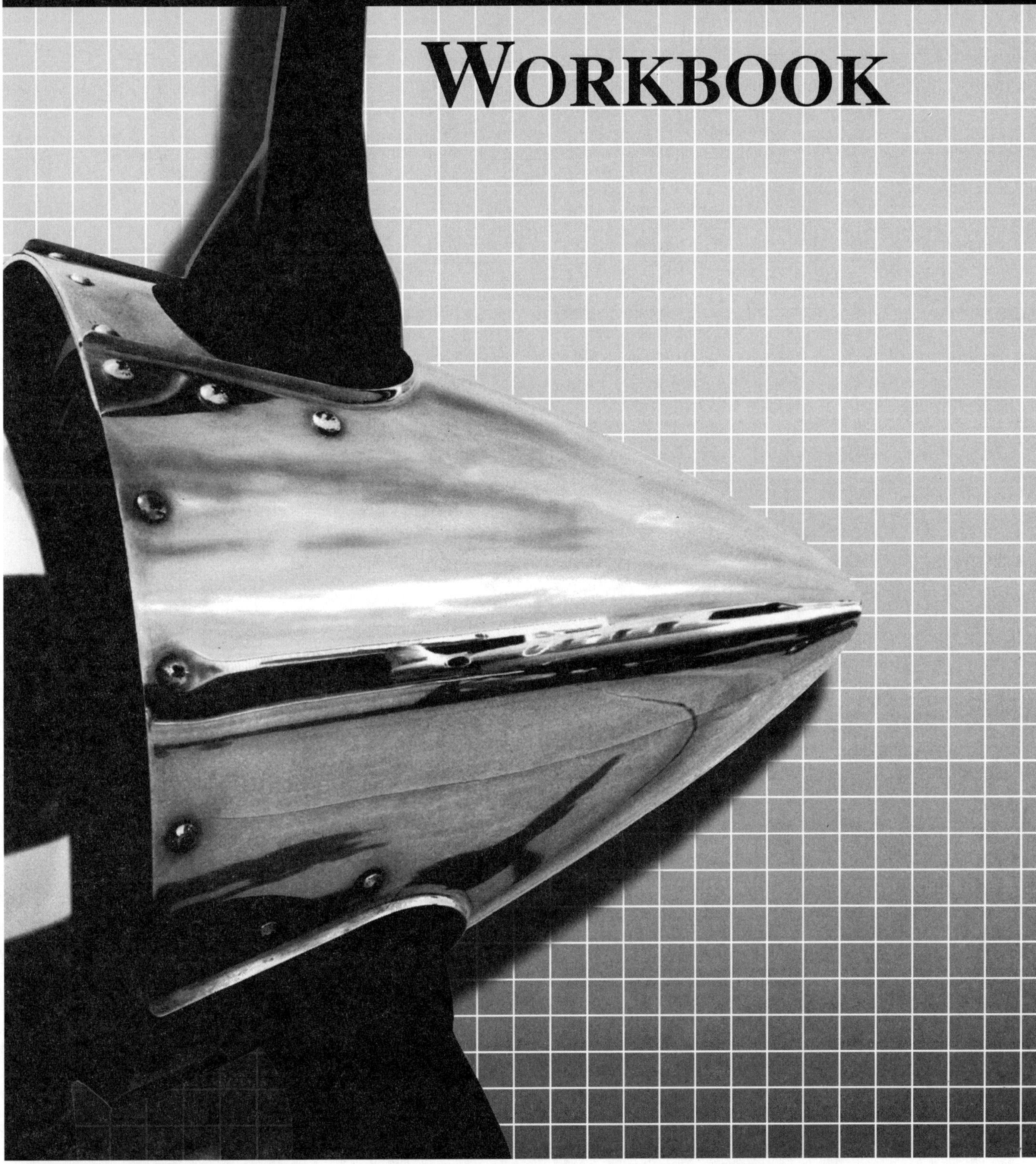

JS322712A

55 Inverness Drive East, Englewood, CO 80112-5498

PREFACE

The *A & P Technician Powerplant Workbook* is designed to help you check your understanding of the material you studied in the A & P Technician Textbook. Each exercise is correlated with a specific textbook chapter and section. The workbook contains multiple choice, true/false, matching, and completion exercises. To answer the multiple choice questions, circle the number of the correct choice. Fill in the appropriate blanks to answer the other questions. Further instructions are provided, when necessary. The answers to all of the exercises are grouped in the Answer Section at the back of the book.

TABLE OF CONTENTS

CHAPTER 1

RECIPROCATING ENGINES

SECTION A
DESIGN AND CONSTRUCTION

1 Radial engines have appeared in two forms: the _______-radial and the _______-radial.

2 The _______ _______-type engine has two banks of cylinders directly opposite each other with a crankshaft in the middle.

3. Horizontally opposed engine crankcases are usually cast from ______________ alloy.

4. Many crankcase halves are sealed using a thin coat of non-hardening gasket compound and a _______ ______.

5. The three main parts of a crankshaft are the

a. ____________________________.

b. ____________________________.

c. ____________________________.

6. The outer surface of the crankpins and journals are hardened by ______________.

7. A crankshaft is checked for _______ (static or dynamic) balance by placing it on knife edges and observing any tendency to rotate.

8. To reduce vibration to a minimum during engine operation _____________ _____________ are incorporated on the crankshaft.

9. What are the three types of bearings in general use?

a. ____________________________.

b. ____________________________.

c. ____________________________.

10. The crankshaft usually rotates in _______ bearings.

11. The _______________ _______ is the link which transmits forces between the piston and the crankshaft.

12. The connecting rod attaches to the _______ _______ of the crankshaft.

13. What are the three types of connecting rod assemblies?

    a. ____________________________.

    b. ____________________________.

    c. ________________________.

14. The ______________-type connecting rod is used on horizontally opposed engines.

15. The ______________ _______ ______________ rod assembly is used in radial engines.

16. The ______________ pins connect the articulating rods to the master rod.

17. Most aircraft pistons are machined from ______________ alloy forgings.

18. Most piston rings are made of high-grade _______ _______.

19. The top rings installed on an aircraft piston are called the ______________ rings.

20. The piston ring below the compression rings is called the _______ ______________ ring.

21. The piston ring at the bottom of the piston skirt is called the _______ ______________ ring.

22. The piston pins used in modern aircraft engines are the _______-______________ type.

23. Another name for a piston pin is a _______ pin.

24. Oil control rings regulate the ____________ of the oil film on the cylinder wall.

25. The cylinder assembly is composed of two major parts:

    a. ____________________________.

    b. ____________________________.

26. A _______ bore cylinder is one where the bore is ground so that the diameter at the top portion of the barrel is slightly smaller than the diameter of the main part of the barrel.

27. The two methods commonly used to harden cylinder walls include:

    a. ____________________________.

    b. ____________________________.

28. Cylinders which have been chrome plated are identified by ______________ (what color) paint.

29. Cylinders which have been nitrided are identified by ______________ (what color) paint.

30. _______ (Nitrided or Chrome plated) cylinders are very susceptible to rust and must be kept covered with a preservative oil.

31. Cylinder heads of air-cooled engines are generally made of ______________ alloy.

32. Heat is transferred away from the cylinder head through ______ ______ cast into the head.

33. There is a greater depth and area of cooling fins on the ______________ (exhaust or intake) side of an air-cooled cylinder head.

34. The propeller shaft end of an aircraft engine is always the _______ (front or rear) end, regardless of how it is mounted in the aircraft.

35. When referring to the right or left side of the engine, always assume you are viewing it from the _______ (front or rear).

36. The heads of poppet valves come in four shapes including:

 a. ______________________________.

 b. ______________________________.

 c. ______________________________.

 d. ______________________________.

37. Hollow exhaust valves may be partially filled with ______________ (what material) to aid in their cooling.

38. Valve _______ is the distance that valve is lifted off its seat.

39. Valve ____________ is the length of time that valve is open.

40. The valve mechanism of a horizontally opposed engine is operated by a cam _______.

41. The camshaft on a horizontally opposed engine rotates at _______ the crankshaft speed.

42. The two types of valve lifters used in aircraft engines include:

 a. ______________________________.

 b. ______________________________.

43. The _______ _______ transmit the movement of the cam follower to the rocker arm.

44. One end of the _______ _______ bears against the push rod and the other against the valve stem.

45. The valve mechanism of a radial engine is operated by one or more cam ____________.

46. A four-lobe cam ring on a nine-cylinder radial engine will turn at _______ the crankshaft speed.

47. Hydraulic valve lifters are also known as _______ _______ lifters.

48. The running valve clearance of hydraulic valve lifters is _______.

49. The direction of rotation between a propeller shaft and the crankshaft in an engine using a planetary gear reduction system is _______ (the same or opposite).

50. An aircraft engine having a planetary gear system using an 80 tooth bell gear driven by the crankshaft and a fixed 62 tooth sun gear will have a gear reduction of _______:1. This is the same as 16:_______.

51. Three types of attachments for a propeller to an aircraft engine crankshaft are:

    a. ______________________________.

    b. ______________________________.

    c. ______________________________.

52. _______-type propeller shafts are used on most modern horizontally opposed aircraft engines.

53. Indicate what is meant by each letter or number in the following engine designations.

    a. R-985

    1. R = ____________________________.

    2. 985 = __________________________.

    b. O300-D

    1. O = ____________________________.

    2. 300 = __________________________.

    3. D = ____________________________.

    c. LTIO-540

    1. L = ____________________________.

    2. T = ____________________________.

    3. I = ____________________________.

    4. O = ____________________________.

    5. 540 = __________________________.

SECTION B

OPERATING PRINCIPLES

1. All heat engines convert _______ energy into _______ energy.

2. The two basic types of heat engines are

    a. ______________________________.

    b. ______________________________.

3. The five events that must take place in an internal combustion engine to convert the chemical energy in fuel into mechanical energy are

    a. ______________________________.

    b. ______________________________.

    c. ______________________________.

    d. ______________________________.

    e. ______________________________.

4. The Otto-cycle of engine operation is a constant ______________ (volume or pressure) cycle.

5. On which stroke of a four-stroke cycle engine does each of these events begin to occur?

   a. Intake valve opens ______________.

   b. Intake valve closes ______________.

   c. Exhaust valve opens ______________.

   d. Exhaust valve closes ______________.

   e. Ignition occurs ______________.

6. Valve overlap occurs at the end of the _______ stroke and the beginning of the _______ stroke.

7. Identify each cylinder of the engines below by placing the letter used on the drawing next to the proper cylinder number.

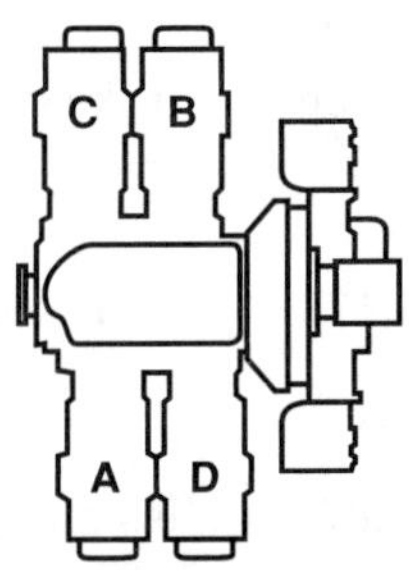

CONTINENTAL FOUR-CYLINDER ENGINE

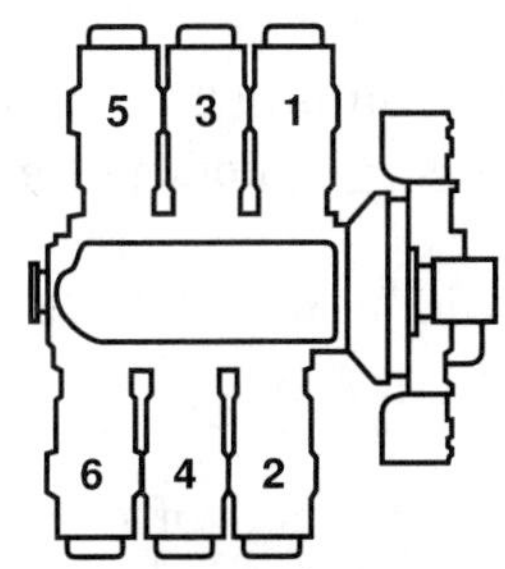

CONTINENTAL SIX-CYLINDER ENGINE

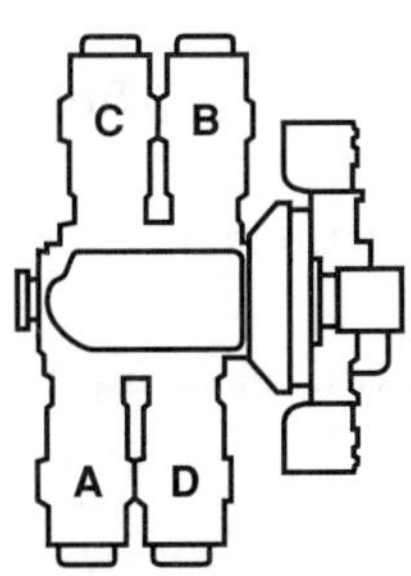

LYCOMING FOUR-CYLINDER ENGINE

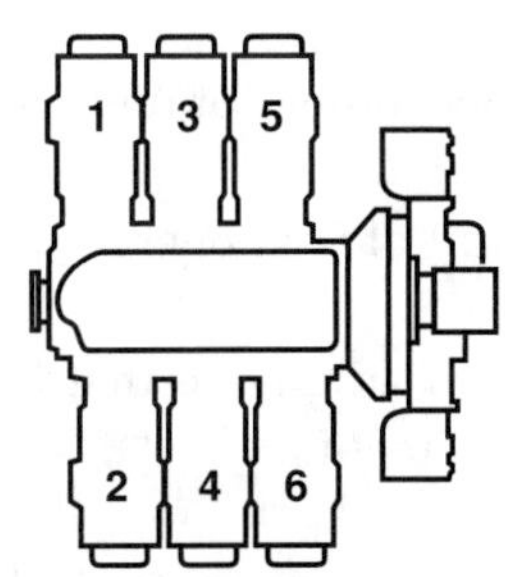

LYCOMING SIX-CYLINDER ENGINE

   a. Lycoming engines #1 ____ #2 ____ #3 ____ #4 ____

   b. Continental engines #1 ____ #2 ____ #3 ____ #4 ____

8. Write the firing order for each of these engines.

   a. Lycoming 4-cylinder horizontally opposed engine ____-____-____-____.

   b. Continental 4-cylinder horizontally opposed engine ____-____-____-____.

   c. Lycoming 6-cylinder horizontally opposed engine ____-____-____-____-____-____.

   d. Continental 6-cylinder horizontally opposed engine ____-____-____-____-____-____.

e. 9-cylinder radial engine ____-____-____-____-____-____-____-____-____.

f. 18-cylinder radial engine ____-____-____-____-____-____-____-____-____-____-____-____-____-____-____-____-____-____.

9. The product of Force × Distance is known as ______________.

10. The rate of doing work is known as ______________.

11. One horsepower is equal to _______ foot-pounds of work done in one minute, or _______ foot-pounds of work done in one second.

12. The horsepower delivered to the propeller shaft of an aircraft engine is called _______ horsepower.

13. The power required to rotate the engine and to compress the air is called _______ horsepower.

14. Brake horsepower + friction horsepower = _______ horsepower.

15. A six-cylinder engine that turns at 2,600 rpm and has a bore of 4.0 inches, a stroke of 4.0 inches, and an IMEP of 120 psi will produce _______ horsepower.

16. A six-cylinder engine with a bore of 4.375 inches and a stroke of 5.125 inches has a piston displacement of _______ cubic inches.

17. A four-cylinder engine with a bore of 4.375 inches and a stroke of 3.875 inches has a piston displacement of _______ cubic inches.

18. The nominal weight of aviation gasoline is _______ pounds per gallon.

19. The nominal heat energy content of one pound of aviation gasoline is _______ Btu.

20. One British Thermal Unit of heat energy will perform _______ foot-pounds of work.

21. The thermal efficiency of an engine that burns 102 gallons of aviation gasoline per hour to produce 1,400 horsepower is _______ percent.

22. A normally aspirated aircraft engine typically_______ (can or cannot) attain a volumetric efficiency of 100%.

23. As the temperature of the air increases, the density ______________ (increases or decreases).

24. The ratio of brake horsepower to indicated horsepower is the _______ efficiency of the engine.

25. ______________ is the spontaneous combustion of the unburned charge ahead of the flame fronts after the ignition of the fuel/air charge in a reciprocating engine.

26. If a cylinder has a volume with the piston at the top of its stroke of 14.5 cubic inches and a piston displacement of 90 cubic inches, its compression ratio is _______:1.

27. If a cylinder has a volume with the piston at the top of its stroke of 23 cubic inches and a piston displacement of 131 cubic inches, its compression ratio is _______:1.

28. _____________ occurs when the fuel/air mixture ignites before the spark jumps across the terminals of the spark plug.

29. A serious loss of power and damage to exhaust valves in an aircraft engine can be caused by ______ (early or late) ignition timing.

30. In order to efficiently produce thrust, the tip speed of a propeller blade must not exceed _____________.

31. The two variables used by a pilot to control the amount of power an engine is producing are engine speed and cylinder pressure, indicated as ______ and ____________ respectively.

32. An engine that develops 180 brake horsepowerwith a brake specific fuel consumption of 0.47 will consume ______ gallons of aviation gasoline per hour.

33. A typical reciprocating aircraft engine converts only about ______% of the heat energy released into useful work.

CHAPTER 2

RECIPROCATING ENGINE OPERATION, MAINTENENCE, INSPECTION, AND OVERHAUL

SECTION A
ENGINE OPERATION, MAINTENANCE, AND INSPECTION

1. By what three means are cockpit controls identified?

   a. ______________.

   b. ______________.

   c. ______________.

2. What color is used to identify the mixture control? _______.

3. The normal range of operation is indicated by a _______ (what color) arc on most instruments.

4. A yellow arc on an engine instrument indicates a _______ range.

5. If the instrument markings are on the cover glass of an instrument a _______ (what color) index mark is required.

6. Fuel pressure gauges are usually calibrated in what units? _______

7. Manifold pressure gauges measure the ______________ (absolute or gauge) pressure inside the intake manifold.

8. Low oil pressure or fluctuating readings on the oil pressure gauge is often a sign of ______________.

9. Engine oil pressure should be indicated within _______ seconds after start-up of an aircraft engine.

10. Before attempting to start any radial engine that has been shut down for more than 30 minutes you should

11. To eliminate a hydraulic lock, remove one of the _______ _______ from each of the lower cylinders and pull the engine through in the direction of rotation.

12. When starting a reciprocating engine, the carburetor heat control should be in the _______ (hot or cold) position.

13. During a ground check or runup, the aircraft should be headed _______ (into or away from) the wind.

14. When performing a power check on an engine with a variable-pitch propeller, the propeller should be in the _______ (low or high) pitch setting.

15. During an idle mixture check if, when the mixture control is moved to the idle-cutoff position, the engine speed does not increase or drops immediately it indicates that the idle mixture is too _______ (rich or lean).

16. A reciprocating engine is stopped by placing the mixture control in the ______________ (full rich or idle cut-off) position.

17. An air leak at the cylinder end of an intake pipe will result in a _______ (rich or lean) mixture within that cylinder.

18. What are the three main parts of an ignition system?

   a. ______________.

   b. ______________.

   c. ______________.

19. Efficient engine troubleshooting is based on a ______________ (random or systematic) analysis of what is happening.

20. ______________ occurs when the contents of the intake manifold or induction system ignite prior to entering the cylinders .

21. An overly rich mixture that continues to burn in the exhaust system is referred to as ______________.

22. The two basic types of compression tests used on reciprocating engines are

   a. ______________.

   b. ______________.

23. The ______________ compression test checks compression by measuring air leakage in the cylinders.

24. The regulator of the differential compression tester should be adjusted to allow _______ psi of compressed air to enter the cylinder.

25. When performing a differential compression test, the cylinder being tested must be on the _______ stroke with the piston at top dead center.

26. For best results, a differential compression test should be performed after the piston rings and cylinder walls have been freshly _______.

27. Engines equipped with ______________ (solid or hydraulic) lifters must have the valve clearances checked and adjusted periodically.

28. Air leaking past the _______ valve during a differential compression test is indicated by a hissing sound heard at the carburetor inlet. Page 2-22

29. A hissing sound heard at the crankcase breather will indicate air leaking past the ______________.

30. When valve clearance is less than it should be, valve overlap is ______________ (lengthened or shortened).

31. A compression value of approximately _______ psi lower than the other cylinders indicates a cylinder leak that must be repaired.

32. When valve clearance is insufficient, the valve will open _______ (early or late) and close _______ (early or late).

33. Before performing a preflight inspection, always be sure that the _______ (ignition or master) switch is in the "OFF" position.

SECTION B
ENGINE REMOVAL AND OVERHAUL

1. The very rapid (in less than one revolution) and complete stoppage of an engine is known as _______________.

2. The regular use of _______________ _______ _______________ provides an operator with a record of metals present in the used engine oil.

3. Engine mount bolts that are to be reused are generally checked using inspection.

4. With a _______________ engine mount, the mounting points are turneb inward so they point toward the engineís center of gravity.

5. The reconditioning of the cylinder, piston, and valve operating mechanism is known as a _______ overhaul.

6. The complete reconditioning of a powerplant is known as a _______ overhaul.

7. _______________ (overhauled or rebuilt) engines are those that have been disassembled, cleaned, inspected, repaired as necessary, reassembled, and tested in accordance with approved standards.

8. _______________ (overhauled or rebuilt) engines are those that have been disassembled, cleaned, inspected, repaired as necessary, reassembled, and tested to the same tolerances and limits as a new item, using either new parts or used parts that conform to new parts tolerances and limits.

9. Hard carbon may be removed by soaking the engine parts in a _______________ solution.

10. Match the following conditions with the term that is used to describe them:

a. _______ A small, round depression in a surface usually caused by the part being struck by another object.

b. _______ A formn of surface corrosion caused by minute movement between two parts clamped together with considerable pressure.

c. _______ A sharp or roughened projection of metal usually resulting from machine processing, drilling, or cutting.

d. _______ A series of deep scratches caused by foreign particles between moving parts.

e. _______ A severe condition of chafing or fretting in which a transfer of metal from one part to another occurs.

f. _______ The breaking loose of small pieces of metal or coated surfaces which is usually caused by defec tive plating or excessive loading.

g. _______ Presence of foreign or extraneous material wholly within a portion of metal. Such material is introduced during the manufacturing process.

h. _______ Wear caused by the rubbing action between two parts under light pressure.

i. _______ A series of blunt depressions in a surface.

j. _______ Loss of metal by a chemical or electrochemical attack.

1. Burr
2. Chafing
3. Corrosion
4. Dent
5. Flaking
6. Fretting
7. Galling
8. Inclusion
9. Peening
10. Scoring

11. Engine parts manufactured from ferrous metals can be inspected for hidden cracks using (dye-penetrant or magnetic particle) inspection.

12. Hydraulic valve lifter plungers and cylinders _______ (are or are not) interchangeable with parts from another assembly.

13. Most engine parts should be tested using both _______________ and _______________ magnetization.

14. When using continuous magnetism, the magnetic particle medium is applied _______________ (at the same time or after) as the magnetizing current.

15. All parts which have been checked using magnetic particle inspection must be ______________ (demagnetized or degreased) before being placed into service.

16. Hydraulic valve lifter plunger assemblies _______ (should or should not) be checked using the magnetic particle inspection.

17. All nonferrous engine parts should be inspected using the ______________ (dye-penetrant or magnetic particle) inspection method.

18. Dye-penetrant may be applied by

    a. ______________.

    b. ______________.

    c. ______________.

19. The time that penetrant is allowed to stay on a part is called the _______ time.

20. The two sets of limits that appear in a manufacturer's Table of Limits are

    a. ______________.

    b. ______________.

21. Connecting rods may be checked for _______ (bend or twist) using arbors and a parallelism gauge.

22. What two measurements should be made when installing new piston rings on the piston?

    a. ______________.

    b. ______________.

23. Dimensional inspection of cylinder barrels should include

    a. ______________________________.

    b. ______________________________.

    c. ______________________________.

    d. ______________________________.

    e. ______________________________.

24. All measurements involving cylinder barrels should be taken at a minimum of two positions _______ degrees apart.

25. Crankshafts are checked for straightness by placing them in vee-blocks and using a _______ indicator.

26. A .003 oversize stud will be identified by the prefix P003 after the part number and _______ (what color) paint on the stud.

27. When the valve face is ground between 1 and 1-1/2 degrees flatter than the valve seat it is referred to as an ______________ fit.

28. Valves which have been ground so there is no margin left are said to have a ______________ edge.

29. Cylinder barrels which are worn beyond limits may be restored by

 a. ______________.

 b. ______________.

 c. ______________.

30. The three grades of stones used in refacing valve seats are

 a. ______________.

 b. ______________.

 c. ______________.

31. The maximum pressure used on a valve seat grinding stone should never be more than that exerted by the weight of the ______________.

32. Valve seating may be checked using ______________ transfer dye.

33. The contact surface of a valve and valve seat should be _______ to _______ the width of the valve face, and in the middle of the face.

34. A 30 degree valve seat may be narrowed using stones ground to an angle of _______ degrees and degrees.

35. Chrome-plated rings may be used with _______ (what material) cylinders.

36. Only _______ _______ (what material) rings can be used with nitrided or chrome-plated cylinders.

37. The end gaps of piston rings should be ______________ (staggered or aligned) on the piston.

38. After an overhaul, an engine should be run-in or _______ tested.

39. What term is used to describe the rigging procedure that ensures the travel of the throttle valve is not limited by the stops on the throttle control quadrant? ______________

40. To prevent failure of the engine bearings during the first start after an overhaul, some form of, _______-______________ is required.

41. One of the steps in preserving an engine is to spray a corrosion preventative compound into each cylinder with the piston at bottom dead center. After the last cylinder is coated with the corrosion preventative compound, the crankshaft ______________ (should or should not) be moved.

42. Cobalt chloride treated _______ gel is used in dehydrator plugs which can be screwed into the spark plug holes of a reciprocating engine.

43. Dehydrator plugs are _______ (what color) when relatively low humidity exists.

CHAPTER 3

TURBINE ENGINES

SECTION A

DESIGN AND CONSTRUCTION

1. The first flight by a jet propelled aircraft took place in ________ (what year).

2. Jet propulsion is a practical application of Newton's ________ (first, second, or third) law of motion.

3. The four types of gas turbine engines are the ________, ________, ________, and the ________.

4 Both turboprop and turboshaft engines must use a ________ ________ between the engine and the propellers/rotors.

5. A gas turbine engine that delivers power through a shaft to operate something other than a propeller is referred to as a ____________ engine.

6 A high bypass ratio turbofan engine has a bypass ratio of ________ to 1 or greater.

7. Engines that use a separate turbine that is not mechanically connected to a compressor to drive a propeller are called ________ ________ engines.

8. Turbofans are generally divided into what three classifications?

   a. ________

   b. ________

   c. ________

9. With modern high bypass engines, most of the engineís thrust is produced by the ________ rather than from the ________ exhaust ________.

10. Bypass ratios of 30:1 are obtainable by what is called an ________ High Bypass (UHB) engine.

11. The inlet duct of a turbine engine is normally considered to be a part of the ________ (airframe or engine).

12. When subsonic air passes through a section of a duct that has a divergent shape, its velocity will ____________ (increase or decrease). Page 3-10

13. Subsonic inlet ducts are usually of a fixed geometry and ____________ (converging or diverging) in shape.

14. Supersonic aircraft typically utilize a ____________-____________ inlet duct.

15. What is the primary function of the compressor section of a gas-turbine engine?
SUPPLY AIR FOR COMBUSTION

16. The two basic types of compressors used on turbine engines are:

a. CENTRIFUGAL

b. AXIAL

17. The three main components of a centrifugal compressor are:

a. IMPELLER

b. DIFFUSER

c. MANIFOLD

18. More than 2 (what number) stages of compression is not practical using a centrifugal compressor.

19. Modern centrifugal compressors can obtain a compression ratio as high as 15 :1.

20. The two main elements of an axial flow compressor are:

a. ROTOR

b. STATOR

21. Axial flow compressors have a pressure rise per stage of approximately 1.25 :1.

22. The rotating elements of an axial flow compressor are referred to as BLADES (blades or vanes).

23. The stationary elements of an axial flow compressor are referred to as VANES (blades or vanes).

24. Multi-spool axial flow compressors have the advantages of:

a. BETTER PERFORMANCE @ HIGH ALTITUDES

b. HIGHER COMP RATIO

c. SMALL FRONTAL

d. HIGH FAN EFFICENCY

25. Compressor blades which have a reduced thickness at the tips are called PROFILE tips.

26. SHROUDED vane segments may be used to minimize vibration.

27. On a multi-spool engine, the low-pressure compressor (N_1) will SPEED UP (speed up or slow down) with an increase in altitude.

28. What two factors determine the angle of attack of a compressor blade?

a. INLET AIR VELOCITY

b. COMPRESSOR RPM

29. Compressor stalls may cause the air flowing in the compressor to SLOW DOWN, or stagnate. In some cases the air may reverse direction.

30. Design factors which contribute to the prevention of compressor stalls include:
    a. VARIABLE INLET GUIDE VEINS
    b. BLEED VALVES
    c. VARIABLE STATOR VEINS

31. The special set of vanes located at the discharge end of the compressor form the outlet vane assembly. The purpose of this assembly is to help straighten the airflow and eliminate any SWIRLING.

32. To take advantage of the good points of both the centrifugal flow and axial flow compressors, some engines may use a HYBRID compressor.

33. High temperature, high pressure air taken from the compressor for various purposes is known as BLEED air.

34. Some current applications of bleed air are:
    a. PRESSURIZATION
    b. HEATING
    c. DE ICING
    d. COOLING
    e. PNEUMATIC START

35. The section of a turbine engine directly behind the compressor is known as the DIFFUSER.

36. The pressure in the DIFFUSER section of a gas turbine engine is the highest of any section in the engine.

37. As primary air flow enters a combustor, it passes through a set of SWIRL VANES, which give the air a radial motion.

38. The three types of combustion chambers that may be found in turbine engines are:
    a. MULTIPLE CAN
    b. CAN ANNULAR
    c. BASKET

39. Annular combustors which reverse the direction of the airflow are REVERSE FLOW combustors.

40. The two types of flameout that can occur in a gas turbine engine are refered to as:
    a. LEAN DIE OUT
    b. RICH BLOW OUT

41. Approximately 60 - 80 (what percentage) of the total pressure energy from the exhaust gasses is absorbed by the turbine section is used to turn the compressor and drive the accessories.

42. What are the two functions of the turbine nozzle?
    a. CONVERT EXHAUST PRESSURE TO VELOCITY ENERGY
    b. DIRECT EXHAUST GAS

43. When turbine blades are installed on a turbine disk, the unit is known as a turbine ______.

44. The most common method of attaching turbine blades to the disk is the ______ ______ method.

45. The three classifications used to describe turbine blades are:

   a. ____________

   b. ____________

   c. ____________

46. Turbine blades may be open or ______ at their tips.

47. The primary air, or the air that enters into the combustion process, amounts to about ______ percent of the total air that flows through a gas turbine engine.

48. The jet nozzle imparts a final ____________ (increase or decrease) in the velocity to the gases.

49. Afterburning can produce as much as a 100% increase in thrust, with fuel flows increasing by ______ to ______ times.

50. Turbofan engines ______ (often or seldom) require noise suppression.

51. Gas turbine engine rotors are usually supported by either ______ or ______ (what type) bearings.

52. The turbine of a turboprop engine extracts up to ______ percent of the engines total power output to drive the propeller.

SECTION B
OPERATING PRINCIPLES

1. The operating cycle of events used by turbine engines is known as the ____________-cycle.

2. The Brayton-cycle is known as a constant ____________ (pressure or volume) cycle.

3. A large turbine engine can operate with a thermal efficiency as high as ______ percent.

4. Warm air is ______ (more or less) dense than cool air.

5. As the temperature of the air at the intake of a turbine engine increases, the amount of thrust produced will ____________ (increase or decrease).

6. To counteract the effects of high intake temperatures, some engines use a ____________ thrust augmentation system.

7. Net thrust of a turbine engine will ____________ (increase or decrease) with an increase in altitude.

8. The net effect of airspeed on thrust is that the thrust ______ (will or will not) increase as airspeed increases.

9. The speed at which the ram air pressure rise cancels the pressure drop inside the inlet duct is known as the RAM RECOVERY point.

10. In most modern engines, power is set using the EPR indicator.

CHAPTER 4

TURBINE ENGINE OPERATION, INSPECTION, MAINTENANCE, AND OVERHAUL

SECTION A
OPERATION, INSPECTION, AND MAINTENANCE

1 Engine pressure ratio is the ratio between readings from the compressor inlet and the _______ stage turbine.

2. Match each of these descriptions with the appropriate turbine engine instrument.

a. _______ Indicates the total engine internal pressure immediately aft of the last turbine stage.

b. _______ Is an indication of the thrust being developed by the engine.

c. _______ Indicates the torque being developed by a turboprop engine.

d. _______ Indicates rpm in percent of maximum

e. _______ Indicates exhaust gas temperature

1. Torquemeter
2. Tachometer
3. EGT gauge
4. Turbine discharge pressure
5. EPR gauge

3. The start of a turbine engine is indicated by a rise in the ______________ _______ ______________.

4. When an engine start produces a temperature that exceeds the limit, it is known as a _______ _______.

5. A hot start inspection typically _______ (does or does not) require engine removal.

6. When an engine has a normal start, but cannot accelerate or reach idle rpm, it is called a _______ _______.

7. If an engine fire occurs during the start of a turbine engine, you should immediately move the fuel control lever to the "off" position and _______ _______.

8. Gas turbine engine thrust may be rated using several different terms, match each description with the appropriate rating.

   a. ______ This is the maximum allowable thrust for takeoff.

   b. ______ This rating is the maximum continuous thrust.

   c. ______ This is the maximum thrust approved for normal climb.

   d. ______ This is the maximum thrust approve for cruising.

   1. Normal rated power

   2. Takeoff power

   3. Maximum cruise power

   4. Maximum continuous power

9. One takeoff and one landing constitutes one ______ ______.

10. A turbine engine is divided into two sections; a ______ and a ______ section.

11. Compressor ______ occurs from ingestion of sand, dirt, dust, and other fine airborne contaminants.

12. The most frequent discrepancy found while inspecting the combustion section is ____________.

13. The inside of a gas turbine engine may be inspected with a ____________ (what instrument) without disassembling the engine.

14. Engine trimming consists of adjusting the engines ______ ______ ______ after replacement or when an engine does not produce maximum thrust.

15. A water wash performed solely to remove salt deposits on compressor components of aircraft operated over or near the ocean is known as ____________.

16. Compressor cleaning using abrasive grit is accomplished while the engine is ____________ (operating or being motored by the starter).

17. The two methods used to remove dirt, salt and corrosion deposits from compressor components are:

    a. ____________

    b. ____________

18. ______ ____________ is the trade name for an EGT and rpm system test unit.

19. Monitoring performance and mechanical systems over time to provide an overview of engine condition is called ______ ______.

20. Within certain limitations, the internal condition of an engine can be evaluated by the ____________ analysis of lubricating oil samples.

SECTION B
ENGINE REMOVAL AND OVERHAUL

1. Typically, a turbine engine _______ (does or does not) have a TBO rating like a reciprocating engine.

2. The most common damage to a compressor is from _______.

3. What do the initials QECA stand for?

4. The overhaul of an engine module is considered to be a _______ (major or minor) repair.

5. Properly cleaned bearings should be dried with a lint-free cloth or allowed to _______dry.

6. Minor damage to engine compressor and fan blades _______ (are or are not) repairable.

7. A partially clogged fuel nozzle causes a condition known as _______ _______.

8. Stress rupture cracks in a turbine blade appear ______________(at right angles to, along) the blade length.

9. Primary creep occurs on a turbine blade ______________(slowly, during the first run, after 100 hours operation).

10. The most common dye penetrant for daylight use is colored _______ (red, green, orange).

11. If a blade having a higher or lower moment weight must be used for replacement, a turbine wheel with an even number of blabes will take at least _______ blades, while a turbine wheel with an odd number of blades will require at least _______blades.

12. Stress rupture cracks on a turbine engine normally run ______________ (perpendicular or parallel) to the length of the blade.

13. Stress rupture cracks and rippling of the trailing edge of turbine blades are usually caused by excessive ______________.

14. Blade _______ (creep or untwist) occurs in both turbine blades and turbine vanes as a result of excessive gas loads on their surfaces.

15. Blade _______ (creep or untwist) is a term used to describe the permanent elongation of turbine blades.

16. A common graphite lead pencil _______ (is or is not) an acceptable marking device for hot section components.

17. Plasma coating is primarily used to build up parts to their _______ diminsions.

18. Compressor blade to engine housing clearances are normally measured with a _______ _______.

19. After a combustion liner has been repaired by welding, it must be _______ _______ to relieve built-up stresses.

20. Turbine blade _______ (cracking or erosion) is the result of the wearing away of metal, either from the gas flow across the surface or the impingement of impurities in the gases on internal components.

21. If a turbine blade is replaced, the new blade must have a _______ _______ that is equal to the old blade.

22. After installing a turbine engine in a helicopter, it may be necessary to check the alignment of the engine to the ______________.

23. When properly adjusted, engine controls will have a small amount of _______ _______ at each end of their travel.

24. The term _______ refers to adjustments that are made to a turbine engine fuel control.

25. When is it usually necessary to trim a turbine engine fuel control?

    a. ______________

    b. ______________

    c. ______________

26. Trim adjustments generally include only the _______ and the _______ speed of the turbine engine.

27. Substances that can absorb moisture from the atmosphere are referred to as ______________.

28. Trimming a turbine engine fuel control allows it to produce _______ _______ _______.

CHAPTER 5

INDUCTION SYSTEMS

SECTION A
RECIPROCATING ENGINES

1. The induction system consists of these four major components:

   a. _______________

   b. _______________

   c. _______________

   d. _______________

2. What three types of air filter systems have been used on aircraft?

   a. _______________

   b. _______________

   c. _______________

3. Reciprocating engine induction systems can be broadly classified as:

   a. _______________

   b. _______________

4. Some modern horizontally opposed engines that use float carburetors have the carburetor mounted directly to the bottom of the engine _______ _______, and the induction pipes pass through the warm oil.

5. Two advantages of passing the fuel/air mixture through the engine oil sump are:

   a. _______________

   b. _______________

6. Air supplied for carburetor heat _______ (is or is not) filtered.

7. The carburetor heat control should always be in the _______ (hot or cold) position when operating on sandy or dusty surfaces.

8. Warm air is _______ (more or less) dense than cool air.

9. Engines equipped with fuel injection systems are not susceptible to carburetor icing, but are equipped with an _______________ air system to allow airflow in case of filter clogging or impact icing.

10. Alternate air doors may be operated _______________ or _______________.

11. Internally driven superchargers are located ______________ (upstream or downstream) of the fuel metering device.

12. Manifold pressure of a turbocharged engine is measured on the ______________ (engine or turbocharger) side of the throttle valve.

13. An aircraft engine whose rated horsepower has been increased by using a supercharger is called a/an ______ engine.

14. An aircraft whose engines are equipped with a two-speed supercharger will use the ______ (high or low) blower position for takeoff.

15. An aircraft engine that can develop its rated sea level horsepower to a given altitude is called a/an ______________ engine.

16. Superchargers which are powered by the energy of the exhaust gases are called ______________.

17. The turbocharger is located ______________ (upstream or downstream) of the fuel metering device.

18. The three sections of a turbocharger are:

 a. ______________

 b. ______________

 c. ______________

19. "The turbocharger shaft is supported in ball bearings." This statement is ______ (true or false).

20. Two important functions of the oil that flows through the center housing of a turbocharger are:

 a. ______________

 b. ______________

21. Causing an engine to produce an excess of manifold pressure is called ______________ the engine.

22. The valve that is used to control the amount of exhaust gases routed to the turbocharger is known as a ______.

23. When the waste gate is fully closed ______ (all or none) of the exhaust gases must pass through the turbocharger.

24. Turbocharger bearings are cooled by a contineous flow of engine ______.

25. Another name for turbocharger discharge pressure is ______ ______ pressure.

26. The waste gate actuator used with a turbocharger system controlled with an absolute pressure controller is ______ (opened or closed) by oil pressure acting on the piston.

27. The capillary tube restrictor is on the ______ (inlet or outlet) oil line inside the waste gate actuator.

28. A ratio controller of a turbocharger system is installed in ______________ (series or parallel) with an absolute pressure controller.

29. The variable automatic pressure controller (VAPC) maintains a constant upper deck pressure for each position of the ______________ valve.

30. The two pressures sensed by the ratio controller are:

    a. ______________

    b. ______________

31. The ratio controller controls the manifold pressure at _______ (high or low) altitudes.

32. Should the pressurer-ratio controller fail, the _______ _______ _______, found in most systems, will prevent overboosting.

33. Engines using turbochargers that are designed to operate from sea level up to their critical altitude are referred to as _______ _______ _______ engines.

34. The three basic components of a sea level boosted turbocharger system are:

    a. ______________

    b. ______________

    c. ______________

35. The density controller regulates "bleed oil" only at the _______ ______________ position.

36. The _______ _______ (what device) on a turbocharger used for pressurizing an aircraft limits the amount of air that can be supplied to the cabin.

37. Rapid throttle movement can cause a temporary overboost known as an ______________.

38. The differential pressure controller reduces the unstable condition known as ______________.

39. Bootstrapping _______ (is or is not) detrimental to engine life.

40. A reciprocating engine in which exhaust-driven turbines are coupled to the engine crankshaft is known as a ______________ system.

SECTION B
TURBINE ENGINES

1. In a ______________ duct air velocity decreases and air pressure ______________.

2. The air entering the first stage of compression on a supersonic airplane must be at a speed below ______________.

3. In flight, the forward motion of the aircraft produces an increase in intake air pressure known as the _______ _______.

4. The inlet for a turbofan is similar in design to that for a turbojet except that it discharges only a portion of its air into the engine, the remainder passes through the _______.

5. Supersonic inlet ducts must be designed to operate in three speed modes:

   a. _______________

   b. _______________

   c. _______________

6. Supersonic aircraft will use a _______________ _______________ (what shape) inlet duct.

7. The inlet duct on a subsonic turbojet powered aircraft is a form of _______________ (converging or diverging) duct.

8. When turbine engines are calibrated on a test stand, a _______________ (what shape) inlet is typically installed.

9. Helicopter operations may require the use of a _______________ _______________ on the inlet of the engine.

CHAPTER 6

EXHAUST SYSTEMS

SECTION A
RECIPROCATING ENGINES

1. The two general types of exhaust systems used on reciprocating aircraft engines are:

   a. ______________

   b. ______________

2. ______________ allow for expansion and contraction of the exhaust system.

3 The purpose of the _______ surrounding the muffler is to capture heat for other systems.

4. Spring-loaded _______ _______ allow for movement without leakage, and slight misalignment of exhaust components.

5. Exhaust ______________ use the velocity of the exiting exhaust gases to produce a venturi effect.

6. Exhaust system parts should never be marked with a _______ pencil.

7. A crack on an exhaust system component will often show as a _______ gray or _______ black streak on the outside of the part.

8. Exhaust system failure almost always creates what hazards?

   a. ______________

   b. ______________

9. Carbon buildup in the waste gate unit of a turbocharged engine is often referred to as _______ deposits.

10. A turbocharger system places a _______(higher or lower) pressure on the exhaust components.

SECTION B
TURBINE ENGINES

1. A greater amount of thrust can be obtained from a turbojet engine if the exhaust gases are discharged from the aircraft at a velocity _______ (higher or lower) than that at the turbine outlet.

2. Subsonic engines will use a ______________ (converging or C-D) jet nozzle.

3. A ______________ (what shape) tail pipe is used on most subsonic aircraft.

4. Supersonic aircraft utilize a ______________ ______________ (what shape) type of tail pipe.

5. Turboprop exhaust gases provide ______________ (increased or very little) additional thrust.

6. The three functions of thrust reversers are:

   a. ______________

   b. ______________

   d. ______________

7. A thrust reverser system can provide about _______% of the total braking force under normal conditions.

8. There are two types of thrust reversers; _______blockage and _______blockage.

9. High bypass turbojets frequently use only _______stream air for reverse thrust.

10. Unmixed exhaust turbofan engines may have both _______ stream and _______ stream reversers.

11. Some high bypass engines may only have _______ (hot or cold) stream reversing.

12. Older engine installations may have _______ _______retrofitted in order to reduce noise to acceptable levels.

13. The noise generated by a turbofan is _______ (greater or less) than that generated by a turbojet.

CHAPTER 7

ENGINE FUEL AND FUEL METERING

SECTION A
FUEL SYSTEMS

1. The partial or complete interruption of fuel flow due to the presence of vapor in fuel lines is called _______.

2. Fuel which vaporizes too readily can cause _______ _______.

3. The primary purpose of a boost pump is to prevent main fuel pump ______________ and system _______.

4. The three basic types of turbine engine fuels are:

   a. ______________

   a. ______________

   b. ______________

5. The main fuel pumps used with turbine engines are generally the ______________ type.

6. A ______________ (what type) boost element may be built into the engine driven fuel pump.

7. The two most common types of filtering elements found in turbine engine fuel systems are:

   a. ______________

   b. ______________

8. What two sources of heat may be used to warm the fuel of a turbine engine?

   a. ______________

   b. ______________

9. How many levels of filtration are normally required in turbine engine fuel systems? _______

10. The fluid used in a water injection system is a mixture of water and ______________.

11. Water injection in a turbine engine is a means of _______ augmentation.

12. The two locations where water may be injected into a turbine engine are:

    a. _______ _______

    b. _______ _______

SECTION B
RECIPROCATING ENGINE FUEL METERING

1. As the velocity of air increases in the throat of a venturi, the pressure of the air will ______________ (increase or decrease).

2. The chemically perfect mixture of air and fuel (stoichiometric mixture) consists of ______ pounds of air for each pound of fuel.

3. A reciprocating engine will develop maximum power with a mixture of approximately ______:1.

4. Carburetors tend to run richer at ______________ (altitude or ground level).

5. Carburetors in which the air passes upward through the unit are called ______________ carburetors.

6. The throttle valve of a float-type carburetor is located between the ______________ and the engine.

7. A float-type carburetor discharges the fuel ______________ (upstream or downstream) of the throttle valve.

8. What are the four basic components of the main metering system of a float carburetor?

   a. ______________

   b. ______________

   c. ______________

   e. ______________

9. What is the main function of the venturi? ______________

10. Air bled into the main metering system ______________ (increases or decreases) fuel density and destroys surface tension.

11. The fuel level in the float chamber is usually kept about ______ inch below the opening in the main discharge nozzle.

12. The ______________ ______________ system determines the ratio of fuel to air in the mixture.

13. The ______ ______________ system is used to shut off the fuel and stop the engine.

14. When a carburetor equipped with a back suction mixture control is in the lean position, the pressure in the float chamber is the same as the pressure at the ______________ ______.

15. The ______________ system supplies extra fuel when the throttle is opened quickly.

16. The power enrichment system is sometimes called an ______________ system.

17. Carburetor controls should be rigged so they have an equal amount of ______-______ at both ends of their travel.

18. A pressure carburetor discharges fuel ______________ (upstream or downstream) of the throttle valve.

19. What two factors make up the air metering force in a pressure carburetor?

    a. _______________

    b. _______________

20. The mixture control in a pressure carburetor varies the fuel/air ratio by controlling the _______ (air or fuel) metering force.

21. A water injection system used on reciprocating engines may also be referred to as _______-injection.

22. Field adjustments to carburetors are generally limited to:

    a. _______________

    b. _______________

23. _______________ (continuous flow or direct injection) fuel injection systems meter the fuel and deliver it to the intake port of each cylinder.

24. Two functions of the flow divider valve during idle conditions are:

    a. _______________

    b. _______________

25. RSA fuel injector nozzles _______ (are or are not) interchangeable between engines and between cylinders.

26. Lengthening the connecting rod between the throttle fuel valve and the throttle air valve of an RSA injection system _______ (richens or leans) the idling fuel/air mixture.

27. The adjustable orifice in a Teledyne-Continental fuel injection pump adjusts the _______ (high or low) unmetered fuel pressure.

28. The relief valve in the Teledyne-Continental fuel injection pump adjusts the _______ (high or low) unmetered fuel pressure.

29. Vapor is removed from the fuel in a Teledyne-Continental fuel injection system in the _______ _______ (what unit).

30. The orifice in a Teledyne-Continental fuel injection pump used with a turbocharged engine is automatically adjusted by an evacuated _______________.

31. The orifice size in a Teledyne-Continental fuel injection pump is varied by the amount of upper deck pressure. This pressure is actually the _______________ _______________ pressure.

32. The manual mixture control in a Teledyne-Continental fuel injection system is a form of _______________ valve.

33. The _______________ _______________ in the fuel control unit limits the maximum amount of fuel that can flow in a Teledyne-Continental fuel injection system during full-throttle, full-rich operating conditions.

34. Two basic functions of the fuel manifold valve in a Teledyne-Continental fuel injection system are:

    a. ____________

    b. ____________

35. A set of C-size nozzles will flow one-half gallon of fuel per hour ______ (more or less) than a set of B-size nozzles.

36. Fuel injector nozzles may be cleaned by soaking them in ____________ or ____________.

SECTION C
TURBINE ENGINE FUEL METERING

1. Two conditions that the fuel control for a turbine engine must prevent are:

    a. ____________

    b. ____________

2. A turbine engine fuel control meters the fuel by ____________ (weight or volume).

3. The three types of turbine fuel currently in use are:

    a. ____________

    b. ____________

    c. ____________

4. The ____________ (what device) in a turbine engine fuel control shuts off all flow of fuel to the nozzles when the engine is to be stopped.

5. The fuel control on a Rolls-Royce RB-211 engine operates on a ____________ (hydromechanical or electronic) schedule until near full engine power.

6. Full Authority Digital Engine Controls (FADEC) replace with electronics most of the ____________ and ____________ elements found in other fuel controls.

7. The two basic types of fuel nozzles used in turbine engines are:

    a. ____________

    b. ____________

8. The type of atomizing nozzle that provides a single spray pattern is known as a ____________ nozzle.

9. The two types of duplex fuel nozzles are:

    a. ____________

    b. ____________

10. The ______________ (what type) fuel nozzle causes the fuel/air mixture to vaporize before it exits the nozzle.

11. Rather than using a flow divider in each fuel nozzle, some engines use a single ______________ valve which acts as a flow divider.

12. Manifold dumping is a procedure which sharply cuts off combustion and also prevents fuel ______________ as a result of residual engine heat.

13. The term used to describe idle speed and maximum thrust adjustments made to a turbine engine fuel control is ______________.

14. Most manufacturers recommend that all fuel control adjustments be made in the ______________ (increase or decrease) direction.

15. If the ______________ (what term) is correct, the fuel control will reach its internal stop before the cockpit quadrant reaches its forward stop.

CHAPTER 8

ELECTRICAL, STARTING, AND IGNITION SYSTEMS

SECTION A
GENERATORS

1. Most light aircraft use only a _______ (AC or DC) power system.

2. The three types of DC generators are:

   a. _______________

   b. _______________

   c. _______________

3. Since the _______________-wound generator has such poor voltage regulation, it is not used as an aircraft generator.

4. Heavy duty DC generators are usually of the _______________-wound type.

5. If the series field of a compound-wound generator aids the shunt field, the generator is said to be _______________-compounded.

6. Compound generators are used where _______ _______________ is of prime importance.

7. The voltage output of a generator is controlled by controlling the strength of the _______ current.

8. Generator speed is generally _______ to _______ times crankshaft speed.

9. If the points on a vibrating voltage regulator stick closed, the generator output will _______ (increase, decrease, or cease).

10. The three units of a three-unit voltage regulator are:

    a. _______ _______

    b. _______ _______

    c. _______ _______ _______

11. The function of the reverse current cut-out is to prevent the generator from _______________ and discharging the battery.

12. Heavy-duty DC generators will use a _______ _______ (what type) voltage regulator.

13. The operation known as ______ ______ ______ is used to restore residual magnetism to the DC generator.

14. New brushes are seated by using ______ (silicon carbide, emery, or sandpaper) wrapped over the armature.

SECTION B
ALTERNATORS

1. The DC alternator produces AC and then rectifies it using six ______ (what component).

2. A three phase alternator has stator windings that are ______ degrees apart.

3. To stop the flow of field current, most systems use a field switch activated by the ______ ______.

4. A DC Alternator ______ (will or will not) have a reverse current cut-out relay in the regulator circuit.

5. The two basic problems that would prevent an alternator from producing electrical power are:

   a. ______

   b. ______

6. If the airplane's battery polarity is accidentally switched, there is a real possibility of damaging the alternators ______.

7. It is ______ (sometimes or never) necessary to flash the field of an alternator.

8. When an ohmmeter forward-biases a diode, the resistance will be ______ (high or low).

9. When an ohmmeter reverse-biases a diode, the resistance will be ______ (high or low).

10. Alternating current has the advantage over direct current in that its voltage can be stepped ______ or ______.

11. The value of alternating current can be changed using a ______ (what device).

12. Large transport catagory aircraft have brushless alternators to reduce ______ at ______.

13. In a brushless alternator, the GCU controls the main generator's output by monitoring and regulating the current in the ______ ______.

14. CSD's, or constant speed drives, are typically driven by a ______ ______.

15. The main purpose of a CSD is to allow the alternator to produce AC power at a constant ______.

SECTION C
MOTORS AND STARTING SYSTEMS

1. The direction of rotation in a DC motor is accomplished by reversing the direction of current flow in either the _______ or the_______ _______.

2. A variable speed motor is either a _______ or a _______ motor.

3. The speed of a variable speed motor is controlled by varying the current in the _______ _______.

4. There are three types of AC motors; they are the _______________, _______________, and _______________ types.

5. Inertia starters crank aircraft engines with the energy that is stored in a _______________.

6. The electric motors for direct-cranking starters are of the _______ (series or shunt) type.

7. A Bendix drive uses a drive pinion which moves back and forth on _______________ splines.

8. Some _______________ _______________ (what manufacturer) engines use a right angle worm drive connected to the starter motor.

9. The direct cranking starter used on high-horsepower reciprocating engines consists of these two basic components:

   a. _______________ _______

   b. _______________ _______

10. On a dual spool axial flow compressor gas turbine engine, the starter will rotate only the _______ (high- or low-) speed compressor.

11. Electric starting systems for gas turbine aircraft are of these two general types:

   a. _______ _______

   b. _______-_______

12. Electrically operated starters used for gas turbine engines are usually of the _______________-type.

13. Air turbine starters are used with many gas turbine engines because of their high power-to-_______ ratio.

14. An air turbine starter uses _______ (low or high) pressure air.

15. Three sources of compressed air for starting an engine equipped with an air turbine starter are:

   a. _______________

   b. _______________

   c. _______________

16. The _______________ _______________ (what type) starter is very similar to a small gas turbine engine.

SECTION D
ELECTRICAL SYSTEM COMPONENTS

1. The wiring between components must be of _______ (stranded or solid) construction.

2. Copper wire use for aircraft electrical system installations must meet MIL-W-_______________ specification.

3. Aluminum wire used for aircraft electrical system installations must meet MIL-W-_______________ specification.

4. The standard insulation strength for electrical wire used in aircraft installations is _______ volts.

5. The smallest gauge aluminum wire recommended for use in aircraft systems is _______ gauge.

6. A _______ (what number) gauge aluminum wire could be used to replace a 4-guage copper wire.

7. When selecting the wire size to use in an electrical system installation, you must consider the maximum _______ the wire will carry and the maximum _______ _______ allowed.

8. Complete this chart for the maximum voltage drop:

SYSTEM VOLTAGE	VOLTAGE DROP ALLOWED	
	CONTINUOUS	INTERMITTENT
14		
28		
115		
200		

9. Using the wire chart, find the proper size copper wire to use for the following conditions:
   a. 50-amp load, 14-volt system, continuous load, routed in free air, 14 feet of wire. _______-gauge wire
   b. 12-amp load, 28-volt system, continuous load, routed in a bundle, 60 feet of wire. _______-gauge wire
   c. 20-amp load, 28-volt system, continuous load, routed in a bundle, 30 feet of wire. _______-gauge wire
   d. 200-amp load, 28-volt system, intermittent load, routed in free air, 40 feet of wire. _______-gauge wire

10. Two basic types of electrical wiring installations are:
    a. _______ _______
    b. _______ _______

11. If a wire bundle is routed through a compartment parallel to a fuel line, the wire bundle should be (above or below) the fuel line.

12. No wire bundle should be closer than _______ (how many) inches from a control cable.

13. Electrical wire bundles are fastened to the aircraft structure with insulated _______ clamps.

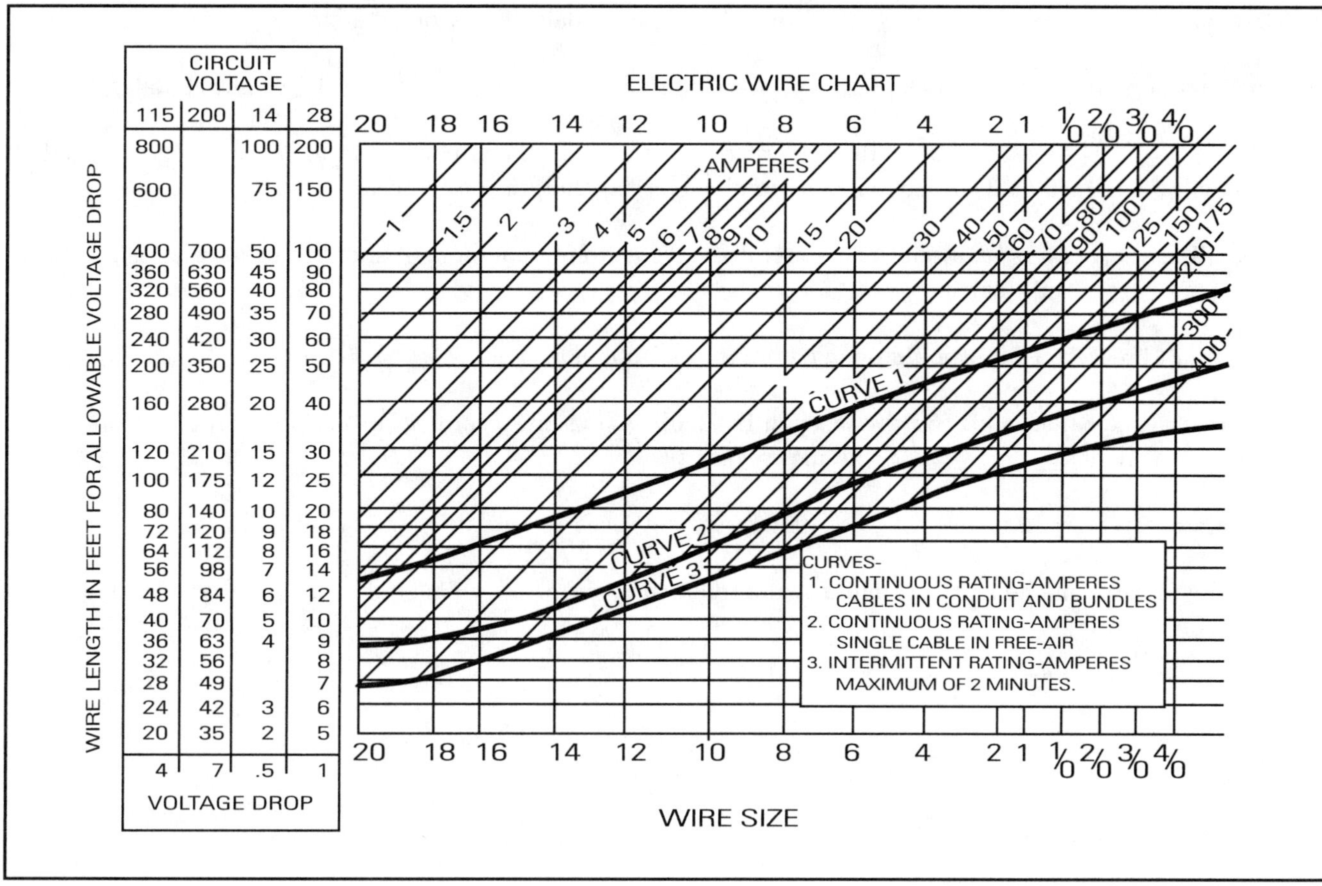

14. The two basic ways to prevent electromagnetic interference in radio and electronic equipment are:

    a. _______________

    b. _______________

15. The inside diameter of a conduit must be about _______ percent larger than the maximum diameter of the wire bundle it encloses.

16. _______________ (bonding or shielding) is used in an aircraft electrical system to intercept radiated electrical energy.

17. Give the color code for the insulation on a solderless terminal used on each of the following size wires:

    a. 16 gauge = _______

    b. 22 gauge = _______

    c. 10 gauge = _______

    d. 18 gauge = _______

18. When a terminal is properly crimped onto an electrical wire, the joint should be _______ (what percent) as strong as the wire itself.

19. The smallest terminal stud allowed for electrical power systems is a number _______.

20. There should never be more than _______ (how many) terminals on any one stud.

21. Most junction boxes in the powerplant area are mounted ______________ (vertically or horizontally).

22. ______________ (Bonding or Shielding) consists of connecting all components in the aircraft together so that there will be no insulated components on which electrical charges can build up.

23. A bonding strap that is required to carry return current from a component should not have a resistance of more than _______ milliohms.

24. The circuit breaker or fuse should open before the conductor _______ _______.

25. A circuit breaker that holds the circuit open if a fault is present, regardless of the position of the operating control, is called a _______-_______ circuit breaker.

26. Automatic reset circuit breakers _______ (are or are not) used in aircraft electrical systems.

27. Give the maximum amount of current a switch rated at 35 amps, 24-volts could carry, in the following types of 24-volt circuits.

    a. Incandescent lamp _______

    b. relay _______

    c. DC motor _______

    d. Heater _______

WIRE SIZE	INSULATION COLOR
16-GAUGE	
22-GAUGE	
10-GAUGE	
18-GAUGE	

28. Two-position "on-off" switches should be mounted so that the "on" position is reached by an (upward or downward) movement.

29. The principal difference between a relay and a solenoid is that a relay has a _______ iron core, while a solenoid has a _______ one.

SECTION E
RECIPROCATING ENGINE IGNITION SYSTEMS

1. Aircraft engines are required to have dual ignition. (true or false) _______

2. A magneto is actually a self-contained _______________ (alternating or direct) current generator.

3. Currently manufactured, certificated, reciprocating engines will use the _______ (high- or low-) tension magneto ignition.

4. What kind of magneto ignition system was developed to prevent high voltage breakdown in the distributor when the aircraft is flying at high altitudes? _______-_______________

5. Dual magnetos contain two separate and independent systems but share the following components:

   a. _______________

   b. _______________

   c. _______________

6. The four basic systems in a rotating-magnet magneto are:

   a. _______________

   b. _______________

   c. _______________

   d. _______________

7. An 18-cylinder radial engine using a low-tension magneto ignition system will have _______ (how many) high-tension transformers.

8. Two methods of mounting a magneto to an aircraft engine are by:

   a. _______________

   b. _______________

9. The pole shoes in a magneto are made of _______________ _______.

10. When the magnet is in its full-register position, the _______ (maximum or minimum) number of lines of magnetic flux are passing through the coil core.

11. The greatest change in the number of lines of magnetic flux in the coil core takes place when the magnet is in its _______________ (neutral or full-register) position.

12. The magnetic field produced by current flowing in the primary coil _______ (opposes or aids) the magnetic field that caused it to flow.

13. Because of the magnetic field produced by the primary current, the greatest change in the total magnetic flux occurs _______ (before or after) the magnet passes its neutral position.

14. The breaker points in a magneto are closed when the rotating magnet is in its ______________ (neutral or full-register) position.

15. The point at which the breaker points open is called the _______ angle.

16. Two functions of a capacitor in a modern magneto are:

   a. To minimize ______________. For this function the capacitor is in parallel with the breaker points.

   b. To minimize ______________ ______________. For this function the capacitor is in series between the brake points and the ignition switch.

17. The high-tension magneto uses a ______ (built-in or separate) distributor.

18. Which winding in a magneto coil has the greater number of turns, the primary or the secondary winding? ______________

19. The distributor in a magneto is in the ______________ (primary or secondary) electrical circuit.

20. The vast majority of four- and six-cylinder horizontally opposed engines use a/an ______________ (what system) to produce a hot and late spark for starting.

21. Magnetos have the problem of not producing a hot enough spark when they turn too _______ (fast or slow).

22. Four methods of providing a hot spark for starting an aircraft engine are:

   a. ______________

   b. ______________

   c. ______________

   d. ______________

23. The Bendix ______________ (what system) is a modern version of the induction vibrator.

24. The extra set of breaker points used with the "Shower of Sparks" system are referred to as the _______ points.

25. On an engine equipped with the "Shower of Sparks" system, the _______ (left or right) magneto is of standard construction.

26. An engine equipped with the "Shower of Sparks" system will use only the _______ (left or right) magneto for starting.

27. The "Shower of Sparks" will jump across the spark plug only as long as which conditions exist? _______ (A, B, or C)

   A. The retard points are open

   B. The advance points are open

   C. Both sets of points are open

28. The rotating magnet in a Slick Electro magneto is in its _______ (neutral, full-register, or E-gap) position when the timing pin can slip through the hole in the housing and the hole in the rotating magnet shaft.

29. You can be sure that the breaker points will close at the correct position of the rotating magnet if their _______ is correct.

30. Magneto-to-engine timing is accomplished using the firing position of cylinder number _______ (which cylinder) as the reference.

31. The impulse coupling _______ (should or should not) be engaged when timing the magneto to the engine.

32. A _______ _______ (what kind of equipment) is used to show exactly the time at which the breaker points open.

33. Close timing of a base-mounted magneto to an aircraft engine is permitted by using a _______ _______.

34. Wear of the breaker points will cause the timing of a magneto to drift _______ (early or late).

35. The engine analyzer is an adaptation of the laboratory _______________ (what instrument).

36. Ignition leads are usually made of _______ _______ or _______ _______ wire, enclosed in a rubber or silicone insulation.

37. Modern ignition harnesses are available with terminal ends to fit either the standard _______ (what size) shielded spark plug or the _______ (what size) all-weather spark plug.

38. The _______________ (capacitance or inductance) caused by shielding of the ignition leads causes excessive electrode erosion around the ignition leads.

39. The _______________ (what component) in a spark plug is installed to minimize electrode erosion caused by using shielded ignition leads.

40. The length of the shell threads on a spark plug is called the _______ of the spark plug.

41. A short reach spark plug has _______-inch of threads on its shell, and a long reach plug has _______-inch of threads.

42. A spark plug that has a long nose core insulator (heat path) is a _______ (hot or cold) spark plug.

43. A _______ (hot or cold) spark plug should be used in a high-compression engine.

44. An engine operating with a too _______ (rich or lean) idle mixture will cause the spark plugs to have black sooty deposits in their firing end cavity.

45. Hard bead-like deposits found in the firing end of the spark plug indicate that the plug has been subject to _______ _______.

46. Lead fouling may occur on a spark plug that is too _______ (hot or cold) for the engine operating conditions.

47. Two sources of oil that can cause spark plug fouling are:

    a. _______________

    b. _______________

48. A spark plug should not be allowed to wear away more than _______ (what fraction) of its electrode dimension before it is discarded.

49. Severe erosion of both the center electrode and the ground electrode may indicate problems.

50. Lead deposits in the firing-end cavity of a spark plug should be removed with a _______________ (what type) cleaning tool.

51. It _______ (is or is not) recommended that the spark plug gap be widened if it has been inadvertently closed too much.

52. The spark plug that is removed from cylinder number one, bottom, should be replaced in cylinder number _______ (what number), _______ (top or bottom), if the firing order of the engine is 1-6-3-2-5-4.

53. Thread lubricant should be placed on the _______ (first or second) thread from the end of the shell when installing a spark plug in an engine.

54. The lead terminal of a spark plug harness should be wiped with a cloth dampend with _______________ or _______________ before it is installed in the spark plug.

SECTION F
TURBINE ENGINE IGNITION SYSTEMS

1. Turbine engine ignition systems operate all of the time the engine is running. This statement is _______ (true or false).

2. Turbine engine ignition systems are rated in _______ (what unit).

3. The current in a typical capacitance-discharge ignition system may be as high as 20 joules and _______ amps.

4. There are usually _______ (how many) igniter plugs in a turbine engine.

5. The two common classifications of turbine engine ignition systems are:

   a. _______________

   b. _______________

6. A low tension system is typically a _______ (28 volt DC or a 115 volt AC) system.

7. A shunted-gap igniter plug is normally used with a _______ (low or high) tension ignition system.

8. The use of alternating current in high tension exciters eliminates the need for a _______ circuit.

9. The gap on a turbine engine igniter plug is _______ (wider or narrower) than that on a reciprocating engine spark plug.

10. Some smaller turbine engines may incorporate a _______ plug-type igniter rather than a spark type.

11. A glow plug is designed to ignite ______ rather than atomized fuel mixtures.

12. An igniter plug ______ (should or should not) be removed immediately after the ignition switch is turned off.

13. Igniter plugs ______ (are or are not) cleaned in the same manner as spark plugs.

14. A troubleshooting test of the igniter plug is to listen for the ______ ______ made when the plug fires.

CHAPTER 9

LUBRICATION SYSTEMS

SECTION A
ENGINE LUBRICATING OILS

1. Five functions of the oil in an engine lubrication system are:

   a. _______________

   b. _______________

   c. _______________

   d. _______________

   e. _______________

2. The resistance of an oil to flow is known as its _______________.

3. Oil that flows freely has a _______ (high or low) viscosity.

4. The oil used in reciprocating engines has a relatively _______ (high or low) viscosity.

5. The measure of the change in viscosity of oil with a given change in temperature is known as the _______________ _______.

6. An engine lubricating oil whose viscosity changes very little for a given change in temperature has a (high or low) viscosity index.

7. The point at which a liquid will begin to give off ignitable vapors is known as the _______ (flash or fire) point.

8. Aviation 80 engine oil has the same viscosity as SAE _______.

9. The use of straight mineral oil in aircraft reciprocating engines is associated with what conditions?

   a. _______________

   b. _______________

10. The lubricating properties of straight mineral oil _______ (are or are not) as good as those of ashless dispersant oil.

11. Synthetic lubricating oils are _______ (multi- or single-) viscosity oils.

12. If incompatible oil has been added to a turbine engine, it is often necessary to _______ and _______ the oil system before refilling with the proper oil.

13. Turbine engine oils _______ (do or do not) use a standard identification system.

14. SAE numbers _______ (are or are not) used to identify turbine oils.

SECTION B
RECIPROCATING ENGINES

1. Two types of lubricating systems for reciprocating aircraft engines are:

   a. _______________

   b. _______________

2. The lubrication is distributed to the parts of a reciprocating engine by these three methods:

   a. _______________

   b. _______________

   c. _______________

3. The main disadvantages of a wet-sump engine are:

   a. _______________

   b. _______________

   c. _______________

4. What two markings must be on the oil filler cap?

   a. _______________

   b. _______________

5. Oil tanks used on engines equipped with Hydromatic propellers will feed the pressure pump from a standpipe, leaving a reserve of oil for _______________ _______________.

6. A gear-type oil pump is a _______ (constant or variable) displacement pump.

7. Constant-displacement oil pumps _______ (do or do not) require a relief valve to control their output pressure.

8. The two types of oil filter systems are:

   a. _______ _______

   b. _______________

9. In the _______ (full-flow or by-pass) filter system, only a portion of the oil is filtered each time it circulates.

10. The most popular filter in common use for general aviation aircraft is the disposable _______ (surface or semi-depth) filter.

11. The Cuno filter is an _______ (edge- or depth-) filtration type oil filter.

12. The _______ _______ (flow control or surge protection) valve determines which of the two possible paths the oil will take through a cooler.

13. When the oil is cold, it flows _____________ (through or around) the core of the oil cooler.

14. _______ _____________ is a system installed on some reciprocating engines to aid in cold weather starting by introducing fuel into the oil system prior to shutdown.

15. Oil pressure is measured at the _______ (inlet or outlet) of the engine driven oil pump.

16. The main contaminants found in used engine oil are:

    a. _____________

    b. _____________

    c. _____________

    d. _____________

    e. _____________

    f. _____________

17. Oil coolers are normally removed and thoroughly cleaned and inspected at _______ _______ (what event).

18. The oil pressure on engines using an adjustable oil pressure relief may be increased by turning the adjustment screw _____________ (clockwise or counter-clockwise).

SECTION C
TURBINE ENGINES

1. Oil consumption in turbine engines is _______ (lower or higher) than that in reciprocating engines.

2. Most turbine engines use _____________ (what type) lubricating oil.

3. Most gas turbine engines for aircraft use a _______ (wet- or dry-) sump oil system.

4. Turbine engine oil tanks must have an expansion space of _______% or 0.5 gallon, whichever is greater.

5. Two methods of determining the amount of oil in a dry-sump system oil tank are:

    a. _____________

    b. _____________

6. Three types of oil pumps used with turbine engines are:

    a. _____________

    b. _____________

    c. _____________

7. The gerotor pump is considered a _______ (positive or variable) displacement pump.

8. One micron = _______ inch.

9. Three types of main system oil filters used with turbine engines are:

   a. _______________

   b. _______________

   c. _______________

10. Fine mesh screens called _______ _______ filters strain the oil just before it is sprayed onto the bearing surfaces of a turbine engine.

11. Check valves in the oil supply lines prevent accumulations of oil in the _______________ _______________ (what area) of the engine while the engine is not operating.

12. The cockpit oil pressure gauge measures the pressure _______________ (upstream or downstream) of the main oil filter.

13. The location of the oil temperature sensor in a turbine engine _______ (does or does not) make a significant difference in readings.

14. If the oil cooler is located in the pressure subsystem of a gas turbine engine, the engine uses a _______ (hot or cold) tank lubrication system.

15. Two types of oil coolers used with aircraft gas turbine engines are:

    a. _______________

    b. _______________

16. If turbine oil is observed to be dark brown or blackish with little or no contaminants present, it is probably caused by a chemical reaction to _______.

17. Oil pressure adjustments on turbine engines are usually made while the engine is _______ (idling or at 75% rated thrust).

18. The scavenge subsystem will have a _______ (greater or less) capacity than the pressure subsystem.

19. Magnetic chip detectors are usually found in the _______________ (pressure or scavenge) subsystem of a gas turbine engine.

20. The Electric Pulsed Chip Detector _______ (can or cannot) discriminate between small wear particles and larger particles.

CHAPTER 10

COOLING SYSTEMS

SECTION A
RECIPROCATING ENGINES

1. A reciprocating engine converts only about _______ (what percent) of the heat energy of the burning fuel into mechanical energy.

2. The radial engine cowling that encloses the engine and provides for pressure cooling is called a/an cowling.

3. _______________ _______________ (what devices) are used between the cylinders of an air cooled engine to force the cooling air to flow between the cooling fins on the cylinders.

4. Cowl flaps on an air-cooled engine are located in such a way that they produce a _______ (low or high) pressure area to increase the airflow through the cylinder fins.

5. Cowl flaps on an air-cooled engine should be _______ (open or closed) during ground runup.

6. The velocity of the exhaust gases creates a _______ (high or low) pressure at the inlet of the augmentor tubes.

7. Accessories and local hot spots are typically cooled with additional air from _______ tubes directed at them.

8. Helicopter engines are typically cooled by a _______ system that provides cooling airflow.

9. The thermocouples used to monitor cylinder heat temperature may be either of these designs:

   a. _______________

   b. _______________

10. If a cylinder head thermocouple lead is too long, you should _______ (cut off or coil and tie) the excess.

11. Cracks in cylinder head fins can sometimes be _______ _______ to prevent further cracking.

12. Dents or bends in cylinder head fins should be _______________ (repaired or left alone) if there is no danger of cracking.

SECTION B
TURBINE ENGINES

1. Nearly all of the cooling air for a turbine engine must be passed _______ (through or around) the engine.

2. The 25% of the airflow through a turbine engine that is used to support combustion is called _______ airflow.

3. Cooling airflow is called _____________ airflow.

4. The coolest point within a gas turbine engine is at the _____________ _______.

5. The engine exterior and the engine nacelle may be cooled by passing _______ between the case and the shell of the nacelle.

6. The thin walls of the combustors of a gas turbine engine are protected from heat by a flow of _______.

7. The hottest point within a gas turbine engine is in the _______ section.

8. Cooling hollow vanes and blades with bleed air is known as _____________ cooling or _______ cooling.

9. Cooling air passing through porous high temperature material is known as _____________ cooling.

10. What may be used to reduce the temperature of the structure in the vicinity of the exhaust duct? _____________ _____________

CHAPTER 11

ENGINE FIRE PROTECTION

SECTION A
FIRE DETECTION SYSTEMS

1. The _______________ (spot detector or thermocouple) sensors operate using a bimetallic thermoswitch that closes when heated to a preset temperature.

2. The _______________ (thermal switch or thermocouple) type of fire detection system operates on a rate-of-rise principle.

3. The thermocouple in an Edison system that is located where it is relatively protected from the initial flame is known as the _______________ thermocouple.

4. The two relays used in the warning circuit of a thermocouple-type system are:

   a. _______________

   b. _______________

5. The continuous loop detection system works on the same basic principle as the _______________ (spot detector or thermocouple) fire warning system.

6. Continuous loop fire detectors are _______________ (overheat or rate-of-rise) systems.

7. The two types of continuous loop systems widely used are:

   a. _______________

   b. _______________

8. The _______ (Kidde or Fenwall) has two wires inside the inconel tube.

9. The two types of pressure-type sensor responder fire detection systems are:

   a. _______________

   b. _______________

10. Carbon monoxide detectors are generally located in aircraft cabins or _______ areas.

11. The _______________ _______________ (what unit) may be used to test certain continuous loop detection systems.

12. Intermittent alarms are most often caused by an intermittent _______ (short or open) in the detector system wiring.

SECTION B
FIRE EXTINGUISHING SYSTEMS

1. The three requirements for a fire are:

   a. ______________

   b. ______________

   c. ______________

2. Identify the class of each fire described below, using the NFPA classification system.

   a. A brake fire on a system using magnesium wheels. _______

   b. A electrical subpanel. _______

   c. A trash can fire. _______

   d. Spilled solvent ignited by a cigarette. _______

   e. An aircraft cabin fire. _______

3. Halogenated hydrocarbon extinguishing agents are identified using a system of _______ numbers.

4. Name the two basic categories into which extinguishing systems are installed:

   a. ______________

   b. ______________

5. Conventional systems are usually associated with the use of ______________ ______________ as the extinguishing agent.

6. The fire extinguishing agent used in most turbine engine HRD fire extinguishing systems are ______________.

7. A blown-out _______ (yellow or red) indicator disk tells the technician that the system has been discharged by an overheat condition at the bottle location.

8. A blown-out _______ (yellow or red) indicator disk tells the technician that the system has been discharged by normal operation of the system.

9. If a cartridge is removed from a discharge valve, it _______ (may or may not) be used in another discharge valve assembly.

10. Fire extinguisher containers are re-weighed at frequent intervals to determine the _______ of _______.

11. Four things that happen when the pilot pulls the fire switch on a Boeing 727 airplane are:

    a. ____________________

    b. ____________________

    c. ____________________

    d. ____________________

12. Care should be exercised when using CO_2 to extinguish a tailpipe fire because of ____________ caused by the chilling effect of the agent.

CHAPTER 12

PROPELLERS

SECTION A
PROPELLER PRINCIPLES

1. The basic function of a propeller on an airplane is to convert engine torque into ________.

2. Propeller blade angle is the angle between the chord of the propeller blade and the ________________ of ________________.

3. The flat surface of a propeller blade is called the ________ (face or back) of the blade.

4. Propeller blade stations are measured, in inches, from the ________ of the ________________.

5. A ________ ________ (what unit) is designed primarily to increase the flow of cooling air to the engine nacelle.

6. The gradual twist in the propeller blade from shank to tip is known as ________ ________________.

7. On a given propeller blade, the blade angle nearest the hub will be the ________ (highest or lowest).

8. The six forces that act on rotating propeller are:

   a. ________________

   b. ________________

   c. ________________

   d. ________________

   e. ________________

   f. ________________

9. Thrust bending force tries to bend the tip of a propeller blade ________________ (forward or rearward).

10. Torque bending force tries to bend the propeller blades in the ________________ (same or opposite) direction of their rotation.

11. Aerodynamic twisting moment tries to move the propeller blades toward a ________ (low or high) pitch angle.

12. Centrifugal twisting moment tries to move the propeller blades toward a ________ (low or high) pitch angle.

13. "Normally, aerodynamic twisting moment is greater than centrifugal twisting moment." This statement is ________ (true or false).

14. Because of vibration problems encountered with certain engine/propeller combinations, the critical range is indicated on the tachometer by a _______ _______.

15. The _____________ (geometric or effective) pitch is the distance, in inches, that a propeller would move forward in a medium that allowed no slip.

16. The _____________ (geometric or effective) pitch is the distance, in inches, that the aircraft actually moves forward in one revolution of the propeller.

17. The difference between geometric pitch and effective pitch is called _____________.

18. _____________ (tractor or pusher) propellers are those mounted on the upstream end of a drive shaft in front of the supporting structure.

19. Automatic propellers are usually termed _____________ _____________ propellers.

SECTION B
FIXED PITCH PROPELLERS

1. An aircraft manufacturer may specify two models of a propeller for the same installation, a _______ and a _______ propeller.

2. _____________-pitch propellers are simple propellers whose blade angle cannot be changed in normal operation.

3. Most wooden propellers are now made of _______ or_______ (what types of wood).

4. Aluminum propeller blanks are _______ (forged or cast) from high strength aluminum alloy.

5. A McCauley propeller designated as 1B90/CM7448 has a diameter of _______ inches and a pitch of inches at the 75% station.

SECTION C
ADJUSTABLE-PITCH PROPELLERS-

1. The propeller governor is an _______ (what variable) sensing device.

2. An aircraft with a constant-speed propeller system will use the _____________ _____________ (what instrument) gauge to indicate the throttle setting, and the _____________ (what instrument) to indicate the propeller setting.

3. The _______ valve directs oil through ports in the drive shaft to or from the propeller to vary the blade angle.

4. The position of the pilot valve is determined by the action of the _____________ (what unit).

5. When the flyweights in a governor tilt outward they ________ (raise or lower) the pilot valve.

6. The action of the flyweights is opposed by a ________________ spring.

7. When the flyweights are tilted outward, the governor is said to be in an ________________ (overspeed or underspeed) condition.

8. If the governor senses an overspeed condition, the pilot valve will direct oil so that the propeller blade angle ________________ (increases or decreases).

9. The two series of McCauley constant-speed propellers are:

   a. ________________

   b. ________________

10. McCauley constant-speed propellers use oil pressure to ________________ (increase or decrease) the blade angle.

11. Hartzell produces two styles of constant-speed propellers, they are:

    a. ________________

    b. ________________

12. The Hamilton-Standard propeller governor is divided into these three parts:

    a. ________________

    b. ________________

    c. ________________

13. The ________________ (compact or steel hub) propeller may be identified by their exposed pitch changing mechanism.

14. Hartzell propeller systems may use either Hartzell or ________________ (what make) propeller governors.

15. The primary purpose of a ________________ (what type) propeller is to eliminate the drag created by a windmilling propeller when an engine fails.

16. The engine ________________ (does or does not) have to be developing power for the propeller to feather.

17. Most multi-engine aircraft are equipped with ________________ (what type) propellers.

18. A feathered propeller has an extremely ________________ (high or low) blade angle.

19. Hartzell compact feathering propellers use ________ ________________ (what force) to move the propeller blades into feather.

20. The feathering function ________________ (can or cannot) override the constant-speed operation at any time.

21. The accumulator used with Hartzell or McCauley feathering propellers is used to assist the propeller ________________ (feathering or unfeathering) in flight.

22. Medium and large reciprocating engine powered transport type aircraft may use the _______ _______ (what make) Hydromatic propeller system.

23. The Hydromatic propeller is made up of three major assemblies, they are:

    a. _______________

    b. _______________

    c. _______________

24. The Hydromatic propeller operates by using _______________ oil pressure on one side of the propeller piston, and _______________ oil pressure on the other side of the piston.

25. An _______________ operated feathering pump is used to supply oil under high pressure to the Hydromatic propeller when the feathering system is actuated.

26. The feathering system of a Hydromatic propeller is actuated by depressing the _______________ button in the cockpit.

SECTION D
TURBOPROP PROPELLERS

1. The blades of a _______________-type propeller can be rotated to a negative angle to produce reverse thrust.

2. The Beta mode of operation of a reversible propeller system is primarily the ________ (ground or flight) operation.

3. A negative torque signal (NTS) provides a signal which _______________ (increases or decreases) propeller blade angle to limit negative shaft torque.

4. If a power loss occurs during takeoff, the thrust sensitive signal of a turbo-propeller system causes the propeller to _______________ (what action) automatically.

5. During ground operation, the ________ (power or speed) lever used with a TPE-331 engine controls the propeller blade angle.

6. During flight operation, the ________ (power or speed) lever used with a TPE-331 engine sets the engine rpm by varying the propeller blade angle.

7. An electric pump is used to _______________ (feather or unfeather) the propeller on a TPE-331 engine.

8. The propeller shaft ________ (is or is not) connected to the gas generator turbine in a PT-6 engine.

9. The propeller governor on an airplane equipped with a PT-6 engine ________ (does or does not) function as a control unit when operating in the Beta mode.

10. Full aft movement of the _______________ (propeller or power) control lever causes the propeller system on a PT-6 engine to go into the feather position.

SECTION E
AUXILIARY PROPELLER SYSTEMS

1. The propeller ______________ system is used to set all propellers at exactly the same rpm.

2. Two undesirable conditions that are minimized by using synchrophasing are:

   a. ______________

   b. ______________

3. Isopropyl alcohol may be used in a fluid ______________ (anti-icing or deicing) system on propellers.

4 Electrical power for propeller deicing is transferred from the aircraft system to the propeller hub through electrical leads which terminate in what units? ______ ______ and ______ ______

5. Electrical deicing systems are usually designed for ______________ (intermittent or continuous) application of power to the heating elements.

SECTION F
PROPELLER INSPECTION, MAINTENANCE, AND INSTALLATION

1. Major repairs and alterations to propellers are normally performed by the manufacturer or a ________.

2. When a wood propeller is stored, place it in a ______________ (horizontal or vertical) position.

3. If a propeller has been subjected to salt water, it should be flushed with _______ _______ (what fluid) until all traces of salt have been removed.

4. Two variables enter into the determination of whether a bent aluminum alloy propeller blade can be straightened. They are:

   a. ______________

   b. ______________

5. The Hydromatic propeller is lubricated by ______________ (what source) operating oil and needs no other lubrication.

6. The two types of propeller unbalance that are possible are:

   a. ______________

   b. ______________

7. The ______________ ______________ ______________ (what tool) may be used to check propeller blade angles while the propeller is on the aircraft.

8. The adjusting screw on the hub cylinder of a compact propeller will change the ________ (high or low) blade angle.

9. The ________ (what term) of a propeller is defined as the path which the tips of the blades follow as they rotate, with the aircraft stationary.

10. The maximum amount the tips of a six-foot metal propeller can be out of track is ________ (what fraction) of an inch.

11. The process of determining the positions of the propeller blades tips relative to each other is known as blade ________________.

12. Propellers whose pitch change mechanisms are oil actuated must be checked with the engine (operating or shut down).

13. The three types of crankshafts used on aircraft engines are:

    a. ________________

    b. ________________

    c. ________________

14. The ________ (what type) crankshaft requires the use of a hub adapter to install the propeller.

15. McCauley constant speed propellers are found only on ________________ (flanged or splined) crankshafts.

18. The hub to taper contact for a propeller hub must be at least ________% contact area.

17. The snap ring inside a propeller hub used on a tapered crankshaft is there to aid in the (installation or removal) of the propeller.

18. The ________ (front or rear) cone used on a splined propeller shaft is made in two pieces.

19. Installing a spacer behind the rear cone on a splined propeller shaft is sometimes used to correct (front or rear) cone bottoming.

20. Propeller controls must be rigged so that moving the control forward will result in a/an (increase or decrease) in rpm.

ANSWERS A

ANSWERS

CHAPTER 1

RECIPROCATING ENGINES

SECTION A

1. rotary, static
2. horizontally opposed
3. aluminum
4. silk thread
5. a. journal
   b. crank pin
   c. crank cheek
6. nitriding
7. static
8. dynamic dampeners
9. a. Plain
   b. Ball
   c. Roller
10. plain
11. connecting rod
12. crank pin
13. a. Plain
    b. Fork and blade
    c. Master and articulated
14. plain
15. master and articulated
16. knuckle
17. aluminum
18. cast iron
19. compression
20. oil control
21. oil scraper
22. full floating
23. wrist
24. thickness
25. a. cylinder head
    b. cylinder barrel
26. choke
27. a. chrome plating
    b. nitriding
28. orange
29. blue
30. Nitrided
31. aluminum
32. coolingfins
33. exhaust
34. front
35. rear
36. a. mushroom
    b. tulip
    c. semi-tulip
    d. flat
37. sodium
38. lift
39. duration
40. shaft
41. 1/2
42. a. solid
    b. hydraulic
43. push rods
44. rocker arm
45. rings
46. 1/8
47. zero lash
48. zero
49. the same
50. 1.775, 9
51. a. tapered
    b. splined
    c. flange
52. Flange
53. a. 1. = radial engine
       2. = 985 cubic inch displacement
    b. 1. = horizontally opposed engine
       2. = 300 cubic inch displacement
       3. = D series
    c. 1. = left-hand rotation
       2. = turbocharged
       3. = fuel injected
       4. = horizontally opposed engine
       5. = 540 cubic inch displacement

SECTION B

1. heat, mechanical
2. a. internal combustion
   b. external combustion
3. a. intake
   b. compression
   c. ignition
   d. power
   e. exhaust
4. volume
5. a. exhaust
   b. compression
   c. power
   d. intake
   e. compression
6. exhaust, intake
7. a. c, a, b, d
   b. b, d, c, a
8. a. 1-3-2-4
   b. 1-4-2-3
   c. 1-4-5-2-3-6
   d. 1-6-3-2-5-4
   e. 1-3-5-7-9-2-4-6-8
   f. 1-12-5-16-9-2-13-6-17-10-3-14-7-18-11-4-15-8
9. work
10. power
11. 33,000, 550
12. brake
13. friction
14. indicated
15. 118.69
16. 462.25
17. 233.02
18. 6
19. 20,000
20. 778
21. 29.10
22. cannot
23. decreases
24. mechanical
25. Detonation
26. 7.21
27. 6.70
28. Preignition
29. late
30. the speed of sound
31. rpm manifold pressure
32. 14.1
33. 30

CHAPTER 2

RECIPROCATING ENGINE OPERATION, MAINTENENCE, INSPECTION, AND OVERHAUL

SECTION A

1. a. position
   b. shape
   c. color
2. Red
3. green
4. precautionary
5. white
6. Pounds per square inch.
7. absolute
8. low oil quantity
9. 30
10. pull the propeller through, in the direction of rotation, a min imum of two complete turns.
11. spark plugs
12. cold
13. into
14. low
15. lean
16. idle cutoff
17. lean
18. a. Magnetos
    b. Ignition Harness
    c. Spark Plugs
19. systematic
20. Backfiring
21. afterfiring
22. a. direct
    b. differential
23. differential
24. 80
25. compression
26. lubricated
27. solid
28. intake
29. rings
30. lengthened
    1.15
32. early, late
33. ignition

SECTION B

1. sudden stoppage
2. spectrometric oil analysis
3. magnetic particle
4. Dynafocal
5. top
6. major
7. Overhauled
8. Rebuilt
9. decarbonizing
10. a. 4
    b. 6
    c. 1
    d. 10
    e. 7
    f. 5
    g. 8
    h. 2
    i. 9
    j. 3
11. magnetic particle
12. are not
13. longitudinal, circular
14. at the same time
15. demagnetized
16. should not
17. dye-penetrant
18. a. brushing
    b. dipping
    c. spraying
19. dwell
20. a. serviceable
    b. new
21. bend
22. a. End gap
    b. Side clearance
23. a. cylinder taper
    b. out-of-roundness
    c. bore diameter
    d. flange warpage
    e. piston skirt to cylinder clearance
24. 90
25. dial
26. red
27. interference
28. feather
29. a. grinding oversize
    b. chrome plateing
    c. replacement
30. a. rough
    b. finishing
    c. polishing
31. gun
32. Prussion Blue
33. 1/3, 2/3
34. 15, 45
35. steel
36. cast iron
37. staggered
38. block
39. Cushion
40. pre-oiling
41. should not
42. silica
43. blue

CHAPTER 3

TURBINE ENGINES

SECTION A

1. 1939
2. third
3. turbojet, turboshaft, turbopro peller, turbofan
4 reduction gearbox
5. turboshaft
6 4
7. free turbine
8. a. Low bypass
   b. Medium bypass
   c. High bypass
9 fan, jet thrust
10. Ultra
11. airframe
12. decrease
13. diverging
14. converging-diverging
15. To supply enough air to sup port combustion and produce thrust
16. a. centrifugal
    b. axial
17. a. impeller
    b. diffuser
    c. manifold
18. 2
19. 15
20. a. rotor
    b. stator
21. 1.25
22. blades
23. vanes
24. a. better performance at high altitudes
    b. ability to obtain higher com pression ratios
    c. small frontal area
    d. high fan efficiency
25. profile
26. Shrouded
27. speed up
28. a. Inlet air velocity
    b. Compressor rpm

29. slow down
30. a. variable inlet guide vanes
    b. bleed valves
    c. variable stator vanes
31. swirling
32. hybrid
33. bleed
34. a. cabin pressurization
    b. heating
    c. cooling
    d. de-iceing and anti-icing
    e. pneumatic engine starting
35. diffuser
36. diffuser
37. swirl vanes
38. a. multiple-can
    b. can-annular
    c. annular or basket type
39. reverse-flow
40. a. lean die-out
    b. rich blow-out
41. 60-80%
42. a. Convert some of the exhaust gases pressure energy into velocity energy
    b. Direct the flow of exhaust gasses to strike the turbine rotor at the correct
43. wheel
44. fir tree
45. a. impulse
    b. reaction
    c. impulse-reaction
46. shrouded
47. 25
48. increase
49. 3, 5
50. seldom
51. ball, roller
52. 85

SECTION B

1. Brayton
2. pressure
3. 50
4. less
5. decrease
6. water injection
7. decrease
8. will
9. ram recovery
10. EPR

CHAPTER 4

TURBINE ENGINE OPERATION, INSPECTION, MAINTENANCE, AND OVERHAUL

SECTION A

1 last
2. a. 4
   b. 5
   c. 1
   d. 2
   e. 3
3. exhaust gas temperature
4. hot start
5. does not
6. hung start
7. continue cranking
8. a. 2
   b. 4
   c. 1
   d. 3
9. flight cycle
10. hot cold
11. erosion
12. cracking
13. borescope
14. fuel control unit
15. desalination
16. operating
17. a. fluid wash
    b. abrasive grit blast
18. Jetcal Analyzer
19. trend analysis
20. spectrometric

SECTION B

1. does not
2. fod
3. Quick Engine Change Assembly
4. major
5. air
6. are
7. hot streaking
8. at right angles to
9. during the first run
10. red
11. two, three
12. perpendicular
13. temperature
14. untwist
15. creep
16. is not
17. original
18. thickness gauge
19. heat treated
20. erosion
21. moment weight
22. transmission
23. spring back
24. trimming
25. a. Engine change
    b. Fuel control change
    c. When the engine is not developing full thrust
26. max, idle
27. desiccants
28. maximum rated thrust

CHAPTER 5

INDUCTION SYSTEMS

SECTION A

1. a. Air scoop
   b. Air filter
   c. Fuel metering device
   d. Intake manifold
2. a. Flocked screen
   b. Paper
   c. Polyurethane foam
3. a. Naturally aspirated
   b. Supercharged
4. oil sump
5. a. Cools the oil
   b. Warms the induction air
6. is not
7. cold
8. less
9. alternate
10. manuallyautomatically
11. downstream
12. engine
13. ground boosted
14. low
15. altitude
16. turbochargers
17. upstream
18. a. Compressor housing
    b. Bearing housing
    c. Turbine housing
19. false
20. a. Lubricate the bearings
    b. Remove heat
21. overboosting

22. waste gate
23. all
24. oil
25. upper deck
26. closed
27. inlet
28. parallel
29. throttle
30. a. Ambient
    b. Manifold air pressure
31. high
32. pressure relief valve
33. sea level boosted
34. a. Exhaust bypass valve
    b. Density controller
    c. Differential pressure controller
35. full throttle
36. sonic venturi
37. overshoot
38. bootstrapping
39. is not
40. turbocompound

SECTION B

1. divergent, increases
2. Mach 1
3. Ram Effect
4. fan
5. a. Subsonic
   b. Transsonic
   c. Supersonic
6. converging diverging
7. diverging
8. bellmouth
9. particle separator

CHAPTER 6 EXHAUST SYSTEMS

SECTION A

1. a. Short stack
   b. Collector
2. Bellows
3 shroud
4. ball joints
5. augmentors
6. lead
7. flat, sooty
8. a. Fire
   b. Carbon monoxide
9. coke
10. higher

SECTION B

1. higher
2. converging
3. covergent
4. converging diverging
5. very little
6. a. Aid in braking and directional control during normal landing and reduce brake maintenance.
   b. Provide braking and directional control during emergency landings and balked takeoffs.
   d. Back an aircraft out of a parking spot.
7. 20
8. mechanical, aerodynamic
9. cold
10. hot, cold
11. cold
12. hush kits
13. less

CHAPTER 7 ENGINE FUEL AND FUEL METERING

SECTION A

1. vapor lock
2. vapor lock
3. cavitationvapor lock
4. a. Jet A
   a. Jet A1
   b. Jet B
5. spur gear
6. centrifugal
7. a. Steel mesh
   b. Cellulose fiber
8. a. Oil
   b. Bleed air
9. 2
10. alcohol
11. thrust
12. a. Compressor inlet
    b. Combustion chamber

SECTION B

1. decrease
2. 15
3. 12
4. altitude
5. updraft
6. venturi
7. upstream
8. a. Venturi
   b. Main metering jet
   c. Main discharge nozzle
   e. Throttle valve
9. Decrease the pressure
10. decreases
11. 1/8
12. mixture control
13. idle cutoff
14. discharge nozzle
15. acceleration
16. economizer
17. spring-back
18. downstream
19. a. Impact air
    b. Venturi suction
20. air
21. anti-detonation
22. a. Idle speed
    b. Idle mixture
23. Continuous flow
24. a. Create downstream pressure for the fuel control
    b. Cuts off the fuel and leaves the lines full of fuel
25. are
26. richens
27. high
28. low
29. fuelpump
30. bellows
31. turbocharger discharge
32. selector
33. metering orifice
34. a. Distribute fuel evenly to all cylinders
    b. Provide a positive shutoff of fuel
35. more
36. Acetone, lacquer thinner

SECTION C

1. a. Lean die-out
   b. Rich blow-out
2. weight
3. a. Hydromechanical
   b. Electronic
   c. Hydro-pneumatic
4. shutoff valve
5. hydromechanical
6. hydromechanical, pneumatic
7. a. Atomizing
   b. Vaporizing

8. simplex
9. a. Single line
   b. Dual line
10. vaporizing
11. pressurizing and dump
12. boiling
13. trimming
14. increase
15. springback

CHAPTER 8

ELECTRICAL, STARTING, AND IGNITION SYSTEMS

SECTION A

1. DC
2. a. Series
   b. Shunt
   c. Compound
3. series
4. compound
5. cumulative
6. voltage regulation
7. field
8. 1 1/8, 1 1/2
9. increase
10. a. Voltage Regulator
    b. Current limiter
    c. Reverse current cutout
11. motoring
12. carbon pile
13. flashing the field
    sandpaper

SECTION B

1. diodes
2. 120
3. master switch
4. will not
5. a. Short or open diode
   b. Open in field circuit
6. diodes
7. never
8. low
9. high
10. up, down
11. transformer
12. arcing, altitude
13. exciter field
14. hydraulic motor
15. frequency

SECTION C

1. armature, field windings
2. shunt, series
3. field windings
4. universal, induction, synchronous
5. flywheel
6. series
7. helical
8. Teledyne Continental
9. a. Motor assembly
   b. Gear section
10. high
11. a. direct cranking
    b. starter-generator
12. starter-generator
13. weight
14. low
15. a. Aircraft mounted auxiliary power unit
    b. Ground power unit
    c. Bleed air from operating engine
16. fuel/air combustion

SECTION D

1. stranded
2. 22759
3. 7072
4. 600
5. 6
6. 2
7. current, voltage drop
8.

NOMINAL SYSTEM VOLTAGE	ALLOWABLE VOLTAGE DROP VOLTS	
	CONTINUOUS OPERATION	INTERMITTANT OPERATION
14	0.5	1.0
28	1.0	2.0
115	4.0	8.0
200	7.0	14.0

9. a. 6
   b. 10
   c. 10
   d. 1/0
10. a. Open wiring
    b. In conduit
11. above
12. 3
13. cushion
14. a. Bonding
    b. Shielding
15. 25
16. Shielding
17. a. blue
    b. red
    c. yellow
    d. red
18. 100%
19. 10
20. 4
21. vertically
22. Bonding
23. 3
24. emits smoke
25. trip-free
26. are not
27. a. 4.37
    b. 8.75
    c. 11.7
    d. 17.5
28. upward
29. fixed, moveable

SECTION E

1. true
2. alternating
3. high-
4. low-tension
5. a. Housing
   b. Rotating magnet
   c. Cam
6. a. Mechanical
   b. Magnetic
   c. Primary electrical
   d. Secondary electrical
7. 36
8. a. Base mount
   b. Flange mount
9. laminated steel
10. maximum
11. neutral
12. opposes
13. after
14. full-register
15. E-gap
16. a. point arcing
    b. electromagnetic radiation
17. built-in
18. Secondary
19. secondary

20. impulse coupling
21. slow
22. a. Booster magneto
    b. Induction vibrator
    c. "Shower of sparks" system
    d. Impulse coupling
23. "Shower of sparks"
24. retard
25. right
26. left
27. C
28. E-gap
29. maximum opening
30. 1
31. should not
32. timing light
33. vernier coupling
34. early
35. oscilloscope
36. stranded copper, stainless steel
37. 5/8-24, 3/4-20
38. capacitance
39. resistor
40. reach
41. 1/2, 13/16
42. hot
43. cold
44. rich
45. lead fouling
46. cold
47. a. Broken piston ring
    b. Worn valve guides
48. 1/2
49. fuel metering
50. vibrator
51. is not
52. 6, top
53. second
54. acetone, M.E.K.

SECTION F

1. false
2. joules
3. 2000
4. 2
5. a. Low tension (voltage)
   b. High tension (voltage)
6. 28 volt DC
7. low
8. vibrator
9. wider
10. glow
11. liquid
12. should not
13. are not
14. snapping noise

CHAPTER 9 LUBRICATION SYSTEMS

SECTION A

1. a. Reduce friction
   b. Provide cooling
   c. Seal and cushion
   d. Protect against corrosion
   e. Clean
2. viscosity
3. low
4. high
5. viscosity index
6. high
7. flash
8. 40
9. a. New/overhauled engines
   b. Older engines
10. are not
11. multi-
12. flush, refill
13. do not
14. are not

SECTION B

1. a. dry-sump
   b. wet-sump
2. a. pressure
   b. splash
   c. Combination pressure/splash
3. a. The oil supply is limited by the oil sump capacity.
   b. Cooling is more difficult.
   c. Oil temperatures are likely to be higher.
4. a. "OIL"
   b. Tank capacity
5. propeller feathering
6. constant
7. do
8. a. full flow
   b. bypass
9. bypass
10. semi-depth
11. edge
12. flow control
13. around
14. Oil dilution
15. outlet
16. a. Gasoline
    b. Moisture
    c. Acids
    d. Dirt
    e. Carbon
    f. Metallic particles
17. engine overhaul
18. clockwise

SECTION C

1. lower
2. synthetic
3. dry-
4. 10
5. a. Sight gauge
   b. Dipstick
6. a. Vane
   b. Gear
   c. Gerotor
7. positive
8. .000039
9. a. Disposable paper or fiber
   b. Cleanable screen
   c. Screen and disc type
10. last chance
11. accessory gearbox
12. downstream
13. does not
14. hot
15. a. fuel/oil
    b. air/oil
16. heat
17. idling
18. greater
19. scavenge
20. can

CHAPTER 10 COOLING SYSTEMS

SECTION A

1. 30
2. NACA
3. Intercylinder baffles
4. low
5. open
6. low
7. blast
8. fan
9. a. Bayonet
   b. Gasket
10. coil and tie
11. stop drilled
12. left alone

SECTION B

1. through
2. primary
3. secondary
4. compressor inlet
5. air
6. air
7. turbine
8. convection, film
9. transpirational
10. Insulation blanket

CHAPTER 11 ENGINE FIRE PROTECTION

SECTION A

1. spot detector
2. thermocouple
3. reference
4. a. Sensitive
   b. Slave
5. spot detector
6. overheat
7. a. Fenwall
   b. Kidde
8. Kidde
9. a. Lindberg
   b. Systron-Donner
10. cargo
11. Jetcal Analyzer
12. short

SECTION B

1. a. Fuel
   b. Heat
   c. Air
2. a. D
   b. C
   c. A
   d. B
   e. A
3. Halon
4. a. Conventional
   b. High rate of discharge
5. carbon dioxide
6. halogenated hydrocarbon
7. red
8. yellow
9. may not
10. state, charge
11. a. Arms the fire extinguisher bottle switch
    b. Disconnects the generator field relay
    c. Shuts off the fuel to the engine, hydraulic fluid to the pump, and bleed air
    d. Deactivates the engine driven hydraulic pump low pressure lights
12. shrinkage

CHAPTER 12 PROPELLERS

SECTION A

1. thrust
2. plane, rotation
4. face
3. centerhub
4. blade cuff
5. pitch distribution
6. highest
7. a. Centrifugal force
   b. Thrust bending force
   c. Torque bending force
   d. Aerodynamic twisting moment
   e. Centrifugal twisting moment
   f. Vibration
8. forward
9. opposite
10. high
11. low
12. false
13. red arc
14. geometric
15. effective
16. slip
17. Tractor
18. constant speed

SECTION B

1. climb, cruise
2. Fixed
3. birch, ash
4. forged
5. 74, 48

SECTION C

1. rpm
2. manifold pressure, tachometer
3. pilot
4. flyweights
5. raise
6. speeder
7. overspeed
8. increases
9. a. Threaded
   b. Threadless
10. increase
11. a. Steel hub
    b. Compact
12. a. Head
    b. Body
    c. Base
13. steel hub
14. Woodward
15. feathering
16. does not
17. feathering
18. high
19. air pressure
20. can
21. unfeathering
22. Hamilton Standard
23. a. Hub, or barrel
    b. Dome
    c. Distributor valve
24. engine, governor
25. electrically
26. feathering

SECTION D

1. reversing
2. ground
3. increases
4. feather
5. power
6. speed
7. unfeather
8. is not
9. does not
10. propeller

SECTION E

1. synchronization
2. a. Noise
   b. Vibration
3. anti-icing
4 Carbon brushes, slip rings
5. intermittent

SECTION F

1. Certified Repair Station
2. horizontal

3. fresh water
4. a. Bend angle
   b. Blade station
5. engine
6. a. Static
   b. Dynamic
7. universal propeller protractor
8. low
9. track
10. 1/16
11. tracking
12. operating
13. a. Tapered
    b. Flange
    c. Spline
14. taper
15. flanged
18. 70
17. removal
18. front
19. front
20. increase